YOU'RE NOT A "DUMB MASS" MONKEY

YOU'RE NOT A "DUMB MASS" MONKEY

BY

MICHAEL W. RHOADS

WAKE UP MAN AND ARISE FROM YOUR SLUMBER!

TATE PUBLISHING
AND ENTERPRISES, LLC

Your not a Dumb Mass Monkey

Published by Tate Publishing & Enterprises, LLC
127 E. Trade Center Terrace | Mustang, Oklahoma 73064 USA
1.888.361.9473 | www.tatepublishing.com

Tate Publishing is committed to excellence in the publishing industry. The company reflects the philosophy established by the founders, based on Psalm 68:11,
"The Lord gave the word and great was the company of those who published it."

Published in the United States of America

ISBN: 978-1-62994-405-0
1. Science/General
2. Philosophy/Religious
13.09.24

PROLOGUE

Oooh, aaahh-Oohh ahhh? Ahhhhh oh ooh ohhh? Aaah ahhh ooh oooh ah! Could you imagine? Ahhhhh ooh ooohh. Ooh, ahhh? As a member of the human race, would it not perturb you to have to hear that all day; every day? Oooh oooh ahhh. Allow that to sink in a moment. Oooh ooh ooh, aah aah aah!

Well that is the language you are accused of speaking Adam; I mean, mankind. Oooh ooh ooh, aah aah aah! It's quite insulting when you really think about it with all your heart, all your mind, all your soul and all your strength.

Well we're going to drop the oooh oooh aah missing link monkey language and venture on a different journey. We're going to stack letter upon letter forming words, forming sentences, forming content, forming a message. It is called speech and is distinctively different than ooh aah. We have the ability to speak, we have the privilege to speak and we bear the responsibility of speaking. Do you know why? We can speak because we're not monkeys.

HELLO TO YOU MY FELLOW MAN
EVERYBODY CLIMB ABOARD
INSTANTLY GRATIFIED TO DISTANTLY SATISFIED
SPECULATIVE THEORIES ARE ADORED
REMEMBER FOREVER BEFORE IT'S TOO LATE
IS A CRY TO MANKIND TO CLEAN THE SLATE
SOAKED WITH THE NOTION OF CREATING OUR FATE
EVEN THE HEATHEN SHOULD SEEK HEAVENS GATE
NEVER FORGET WHAT MESSIAH'S BLOOD HAS RESTORED

א

ת

Forward

It all started when I was but a boy, on the precipice of lifes' offerings, that I became aware of the notion of evolution. I hated it without the need to ponder, analyze, philosophize or even consider its' validity! I knew beyond any shadow of any doubt that this was and is a false teaching; a puffing up of mans' ego and a tearing down of Yahweh's Holy Word!!!

Instinctively, I discerned error throughout this blasphemous "teaching" and promptly jumped into action. I brought this "teaching", bestowed upon all children once High school-*light* is reached, to the attention of my youth pastor as a topic of utmost importance for my generation! I thought this would get the attention of us all and we would stand together as the body of Christ and debunk this vile elevation of mans' pride!! Boy was I in for a big surprise! The things a twelve year-old thinks he knows.

I had no idea that nobody really cared! I was in utter dismay when our youth pastor took up the counter point to evolution and was met, in a church setting no less, with yawns! And discontent? How could these so called "children of God" not care one iota that we were being taught, nay forced to believe, trained, that we are nothing more than evolved monkeys?

I thought this to be an egregious trespass against our Creators' **Word**, and thought it was a pretty simple fight to take up. I mean come on, we're monkeys brought about by simple selection? We are reduced to an ancestry of apes through slime metamorphosis?

I, mistakenly, thought this would make people mad or at the very least open up debate on the matter. I had no idea how thoroughly trained and manipulated we had become, for I was not yet aware of evolutions' two equally effective relatives, psychology and liberal theology. I naively figured "we're all believers in agreement" when it comes to thwarting the attack of our Lord's sovereignty; again, boy was I in error!

What I found is that I stood alone in my angst; I apparently was alone in finding this lie (evolution) offensive. But being only 12 years of age, how was little old me supposed to defend God's Word, or even myself, against the big bad school district?

How was I to stand against this machine of academia? Needless to say I was flabbergasted and utterly baffled so I did what any young lad in my position would do; I acquiesced and belittled the argument as well, then I quit the fight. A position I have unknowingly regretted ever since! But alas!! Praise Yahshua Jesus; here I stand ready, armed and able to pick up where I left off 28 years ago, so here we go. I'm not a "dumb mass" monkey with a missing link buried somewhere back in my heritage.

I'm not a "dumb mass" monkey lacking abilities and skills. I'm not a "dumb mass" monkey who can't think for himself. I'm not a "dumb mass" monkey; and neither are you!

This compilation of thoughts goes out to those on all sides of all isles, which, yes, does include the man I see when I look into the mirror; those complacent, those content, the staunch, the stiff-necked, the "grass is always greener" bunch and the eggshell walking "fence sitters"; we are all included in this diatribe. Buckle up man it will get bumpy, will be harsh and with no apologies; will be politically **in** correct!

I discern the ability to manipulate God's creation is the real culprit that the theory of evolution looks to cover up. That is what this writing is all about: our creation manipulation capabilities, the mis-guided teaching (training) of origins to deflect attention away from the manipulation and the psychology necessary to effectively achieve this manipulative goal. "Slight of hand", "misdirection", "changing focus"- call it whatever you want, it is still the deceiving of man through theoretical truths.

And I also surmise; modern theology, liberally doled out on a weekly basis to an unknowing, unsuspecting body of believers, has severely damaged and compromised the strength and discernment necessary to even detect that psychology and evolutionistic thinking has infiltrated the church.

Just bear this in mind as you tread through this; Christ never doubted his origins, never sought the advice of a psychologist and contrary to modern interpretation of scripture, warned against false teachings for the itchy eared; so neither should we who believe he is the resurrected Messiah believe or seek such things!

Chapter 1: Theoretical Truth

Have you ever wondered why this theory of evolution is taught as fact? Why would man be so desperate to formulate, conjecture then distribute this "opinion" upon us unsuspecting "dumb masses"? Think about it, why would man so desperately yearn to mock his maker? If evolution were stipulated to be a theory, **every time**, I probably would have looked at it differently than I do, but it hasn't been presented as such! Man has taken Darwin's' **theory** and turned it into **truth**, something that super-cedes the **Word** while soothing the **world**! Right off the bat "The fool says in his heart, there is no God" (PSALMS 14: 1 and 53: 1). Man's opinion, his theoretical truth, which is really just a stealth denial of God, grieves (EPHESIANS 4: 30) and blasphemes (MATTHEW 12: 32) the Holy Spirit!

Do you, fellow man, think that this too, along with every other fallacy caressed by man, will go unpunished? "He who is pregnant with evil and conceives trouble gives birth to disillusionment" (PSALMS 7: 14). Are you so egotistical as to think you will escape God's wrath? No, I say it is stored up for those who mock God, for it is written "Do not be deceived: God cannot be mocked. A man reaps what he sows" (GALATIANS 6: 7). Evolution, ha!

Mock through **epalt**, which I will explain much later, leads to outright idolatry: uncensored, unfettered and unlimited idolatry! I also think it necessary to note that the **Word** teaches, while the **world** trains. The **Word** teaches us to know what we need; the **world** trains us to need to "know"! The following writ will point out the sowing, who is doing the sowing and display what's being reaped, by all.

Why on God's green earth would man perpetrate such vile, anti-God thinking and worse, train our children to do so? What's the catch? What's the hidden agenda here? Why the big lie? Ahhh the big fat lies! Here it is in a nutshell-while having a form of godliness-**man hates God**!! Fallen man, in utter depravity, now thinks *he* is a *god* himself i.e. his passion for his "creation", modern technology; therefore man despises, loathes and hates the thought of bowing to a maker in any way! PRIDE!

So, "in lieu of" God, man has "created" all of this stuff at his fingertips, at our beckon call, which is obedient to its' programming! Everything that surrounds us every day on every side is "manmade", fabricated, forged and manufactured for mass production and mass consumption; in essence, the by-product of our training, cloaked as technology, then consumed by the "dumb masses"!

Poly this, plastic that, synthetic this, hybrid that! Here chemical compounds, there elemental components; it's our mixing, melding, smelting, blending, combining and altering of creations' available substances that blinds us. Our ability to manipulate resources elevates our egos and thinking to statuses higher than is due. However, creation manipulation breeds a flood of sorrow inevitably, as you will see.

Forget the fact that all we've done is nibble from the forbidden tree and called it our own invention! You see, man can create nothing from scratch; he has to steal from God and tag it as his very own. Oh we can replicate and duplicate, but actually create? I chuckle!

I muse when I look at it through proper perspective. What is the proper perspective? God's unblemished Word! What's the amusement? The fact that man would struggle and strive to stand as a *god* is written, prophesied, and taught in the very Word that evolutionists attempt to ignorantly negate or purposely destroy!

The very Word that states this would happen is the same Word man tries so desperately to disprove and that's ironically funny to me! The fruits of this ever-dangerous training serve none other than man's desires! Again, just as Gods' Word teaches it would! All of this I find-bite my tongue- laughable and I take it personally!

I take it personally because many years ago evolution left with me a foul sense, and erroneous view, of everything scientific. I wrongly equated science with evolution thus missing out on the nugget of knowledge that science is **in fact** of God, not opposed to him as man trains us to believe! Man has repeatedly taken science and run with it to and fro in an attempt to discredit God! At 40 yrs. old this is a discernible topic for me to tackle but at 12 this was a nearly fatal blow to my notion of God, creation and life!

This false notion led me on a long 28-year journey away from the scientific aspect of our heavenly, awesome Creator! I already knew, as a child should, that all origins and genesis of life were of, by, through and for Gods' glory! A child knows this instinctively and has to be trained to believe otherwise for mans' theories to thrive.

My novice sight of God coupled with mans' definition of science drove me away from my God-given right to know God through scientific eyes, as the writers of scripture knew him. For now, but not forever, science is perched opposite the glory of God, deceptively placed there by satan via man.

It stands in contrast to El Shaddi's Holy Word, instead of its' rightful place as a work *of* him!! You see people, science is as everything else visible and measurable to us; it (science) is of God!! It is of *his* infinite knowledge and given birth through *his* all-powerful, all knowing, and all encompassing Wisdom.

Science is not the creation of man; it is but a stolen morsel from the Wisdom of God by way of sorcery, witchcraft and magic!! Evolution, however, is the doing of man: the pollution of all notions of beginnings!!! Man- (and I take for granted the inclusion of woman in this one word, man), -in his quest for knowledge-*ever heard that word before? Maybe plucked from a tree in a place called Eden*? -has intentionally attempted to puff himself to the loftiest of heights so as to attain that aha! -particle collider moment.

A "gottcha" moment if you will, the proof to lend credence to his anti-God and sadly, through extension, anti-Christ theories! We lay claims of ownership to things not ours, as we steal from God and call it our own. Then we abuse it, turn, and blame God? Heinous!!

Letters, words, language, speech, numbers, equations, calculation and math were all endowed upon us by our Creator; they're not evolved traits from a chaotic cosmic accident! It takes more faith, the thing burped all over by those in the "know", to believe this all came to be by accident than to take God at his Word and then believe and thank God for his Word!

There is something of special note that I think many in today's society take for granted. God didn't have to give us the written Word and he still would have remained right, just, loving and all-powerful! He did it for us!!! It, his Word and all creation for that matter, was and is a gift to and for us! Get it yet, people? You see God wants us to take him at his Word, period!

If, however, you don't want to believe God's Word, even though it is written in more places than just scripture, then we must ideologically part ways my fellow man; a proverbial wiping clean of the hands, if you will! However, for those of you who feel at least slightly ostracized by the magician laden, scientific, in the "know" crowd; let's learn about the attributes of God.

For the fence occupiers who reside in the middle, tip-toeing on eggshells because of the fear of upsetting the apple cart, read REVELATION 3: 15-16. Although it may "feel" safe, this modern area of politically correct gray is not a good place to be found, according to our Creator!

Here is the truth and heart of the matter; both numbers and letters are of, yes from, the spiritual realm! All matters of calculation and speech are what separates us and sets us above the creatures of the sea, the beasts of the field and the birds of the air, (GENESIS 1: 26)! And yes, even the monkeys.

We were essentially placed above everything else that houses, within its' flesh, the breath (Spirit) of our great Creator, YHWH Elohim! I think it pertinent to mention the one thing man was never intended to have dominion over…his fellow man! The ability to speak a single word emanates from God so put away your pride! The ability to calculate the simplest of equations, (1+1/1-1), also emanates from our sovereign Lord so enough boasting, man!!

From the beginning we were given the gift of his image stamped upon us! And for those who truly are sons and daughters of the Most High Living God, through the name and blood of Yahshua/Jesus, that stamped image (speech and calculation) is nurtured as insight and Wisdom grows. I'm speaking here of Holy Spirit Wisdom of course not mans' earthly wisdom!

The best of mans' knowledge is as a "filthy" rag to God!! May his Holy Spirit encourage and expound upon you these truths, since they are declared in scripture! Actually, scripture is much harsher when describing the supposed "righteous acts" of man! Man now believes his meandering in In vitro fertilization, stem cell research, trans-humanism, genetic modification and cloning to be “righteous acts” of creating, extending or repairing life, “for the betterment of mankind”; not at all aware of or completely disregarding what was occurring in Noah’s day, and an outright denial of that ages’ end result, all because of denying the bible’s authenticity.

Don’t get me wrong, I find it astonishing what man has learned to do on the atomic, sub-atomic, particle, cellular, molecular and micro-biological levels as well as the next guy! I find science astonishing; I just simply abhor the piously arrogant and Godless attitude that accompanies such “righteous act” accomplishments, through our God-given ability to manipulate God-given resources.

For atheists, as you find yourself in the position of doubting the validity of my argument, I offer you this simple test: take everything-and I do mean everything-that is reached through man adding, subtracting, multiplying or dividing and remove them and tell me what you are left with? Rescind mans’ God-given ability to calculate and deduce for me a remainder!! Outside of Gods’ systems already in place when God placed man here nothing remains.

Now take everything spoken, which would include all written texts, and remove them as well. What do you find? Rescind all manners of speech and what remains? Same scenario! You find you can't do these two things, 1) remove speech and 2) remove calculation and have anything pertinent to man remain!! That is because they, letters and numbers, are of the spiritual realms; and we, mankind, are created to house and utilize these attributes.

They are the fundamental building blocks of life, the method by which our Creator created! Unfortunately for us these building blocks are only viewed through our skewed and mortally fallen nature. Knowledge we have a-plenty, Wisdom we lack severely! Keep in mind the penalty the tree nibbling caused as we forfeited Wisdom to gain knowledge!

Bear in mind that every word we speak is composed one letter at a time and both letters and numbers, when built upon, form words and equations. And the *ability* to calculate and speak is from our Creator, not we the created!

God stretched out the heavens and set the foundations of the earth -with numbers-and then spoke the letters, through words, "Let there be light"! This introduction of numbers and letters by God is precisely when and how dimensions were created, man! Those in the scientific field know this incorporates both invisible and visible light- the light spectrum-, inaudible, and audible sound- the spectrum of sound waves and vibration (frequencies), and multiple dimensions.

Add gravity to the mix and presto, existing comes in to existence! Thus everything from the smallest particle to the largest quasar, event horizon, nova, super nova, pulsar inducing star finds its' origins in, and is measured in, light waves and sound waves; and held harmoniously intact through the vibration of God's Holy speaking; Yahshua Jesus is the Word and voice of God!

This is a revelation only recently "discovered" and divulged by man, yet known from the foundation of everything by our Creator from Everlasting to Everlasting! Do not confuse "Let there be light" on day one with the creation of the sun, moon and stars-"governing lights in the sky"- on day four either! (These are two different aspects of the same creation).

And yes science, when utilized correctly, verifies this. There is the light the sun produces and then there is **The light** found on the electro magnetic, sub-atomic molecular level, which sustains everything known to man. Oh if man could only resist the temptation to sprinkle speculation abroad! Anti matter, dark matter, dark energy, black holes, quantum mechanics, quantum gravity- when will it end?

Today's truth is that you have been lied to my brethren! You have been taught and trained to believe that you came from a monkey when God's Word, and all creation, states you were made in God's image! From the time we reach Jr. High school we are trained to believe that we are descended from a monkey and ultimately, we are the result of an accidental "big bang"!

Not only is the theory part omitted in this training, we are to believe this as truth-no questions asked! Teaching evolution vs. intelligent design is not even allowed in most school districts presently. Atheism, I've heard in a recent poll, makes up only around 3-6 % of our populous here in the U.S., yet their side of the debate is the lone voice taught to our children via our educational systems, why? Why is the evolution-based scientist in fear of opposing voices?

Why the narrow-minded approach? This narrow-mindedness is a topic I will touch on more later, but for now let's just simply ask the questions. What the heck are they afraid of? Why the one sided output of information? What is really at stake here? Why the conspiracy? Why indoctrination? Perhaps they conspire and indoctrinate to hide the simple Truth? Exactly!!

Not only is Christ denied as Christ, God too is denied as our Creator! Blasphemy! Personally, I want no parts of this training nor do I desire to ever bite of its' fruit. A lie is a lie is a lie! As I alluded to earlier: if you, oh man, want to believe a lie and train others to do the same, I shake the dust off my shoes on the matter and I wipe my hands clean!

However, I became extremely offended, disgusted and utterly indignant when forced by law to allow my children to be hammered with this outright lie!! Who are you, oh man? What in your sorcery reached science sets you above the rest of us? Where were you when he, God, laid the foundations of the earth? Tell me if you know! Why does his creation perplex you?

Why so vexed not knowing where and how gravity “evolved”? Your blatant ignorance reeks of folly! As was with JOB, there will be a price to pay for attempting to sequester God! I suggest you return to the Word of him who made you and I highly suggest you take a peek at the book of JOB starting with chapter 38.

“Then the Lord answered out of the storm. He said: ‘Who is this that darkens my counsel with words without knowledge? Brace yourself like a man; I will question you, and you shall answer me’” (JOB 38: 1-7). It would absolutely behoove you as well to read the chapters and verses leading up to and which follow this response from God to gain its’ full context. If you reject my plea then stick out your neck for here comes your millstone, for you are leading the little ones astray! I should know, I’m a grown up little one you led astray and I’m here to call you out!

Remember what Messiah spoke about those who lead the little ones astray? Please, at the very least consider your ways, as there is nothing left for the proud but to be humbled.

"Pride goes before destruction, a haughty spirit before a fall" (PROVERBS 16: 18).

If not, and you just must continue to "have it all figured out" fine, just leave my children out of it!! My prayer is that you repent and turn and be healed, but my hoping well for you probably just insults you, so I'll leave it at that for now.

For the sake of our children, the future "dumb masses", perhaps it is time for those on both sides of the isle, science and religion, to lay down their "arms" and hear the Truth! We are talking about *only* two dimensions, religion and science, of the same Creator here. They both speak of our all mighty, all knowing, gracious, loving, forgiving and approachable maker, thanks be to our risen Christ! Praise Yahweh, science is his! Praise Yahweh, religion is his! Both speech and calculation is his!

They are ours to toil and contend with, but not ours to claim ownership of! We do not, in any way, own these pillars of creation-we are subject to them. Science and religion were not born of our intellect; they are aspects of our being, strategically placed there by the one who formed us in to being.

And these gifts to us are *only* two of the many attributes, aspects, traits, characteristics and dimensions our phenomenal heavenly Creator possesses! How many more heavenly features did we reject in the garden that we are not yet aware of?

Remember, we ate of a tree we were forbidden to eat of and brought this all on ourselves! Knowledge of good and knowledge of evil came about all because we "bit off more than we could chew", which was accomplished with just one bite! One disobedient act of defiance has left us in utter peril, lest we forget how far man is actually fallen according to YaH.

Read Torah, the first five books in the Old Testament, and realize just how impossible it is, in our fallen state, to please our maker! Then you will understand completely that his perfect laws, creeds, precepts and statutes cannot be met by you, oh fallen man!! We simply cannot comply, thus the Passover at the stake and the Glory of the empty tomb, hence, the Gospel. Only after Christ calls you can you begin to obey Torah!

Redemption is made available through the resurrected Christ, and yes only him! But not for those who would deny the existence of God himself, all in the name of the *religion* of science! Remember this when your pride prevails; even the demons believe in God---"and shudder" (JAMES 2: 19)!

"Intelligent design" which replaced "creationism" and became the new catchword of the debate, I would think, is an insult to God too! We've been arrogantly reduced to groveling at the school districts of America's educational system, just to have the right to have "intelligent design" taught alongside evolution! Of course court ruling after court ruling has denied this petition.

The progressively atheistic science soaked folk fear the Word so they weep and gnarl and gnash to get laws passed attempting to render the Word obsolete. Impressive! However, I would be more impressed with the "scientific community" would they take up this argument head-on! Not: eliminate any voice of dissent, call it a victory and then teach it, through rigorous training, as truth! It seems cowardly, sneaky, covert and shady to me! Put God back in our schools, our courtrooms, and our government and our workplace and *then* teach your "theory"!

Teach them, science and numbers-numbers and scripture, side by side and observe the fruits they bear!! The Truth is that they are both right, just not as man interprets them! You can't teach one as truth, which is mere theory and omit one as false, without proof. That more resembles the training of a monkey than the teaching of the Truth! Which has been the plan all along?

I believe both to be right and in complete harmony when married correctly (I'm talking about the union of **science** and **scripture** here, not "evolution" and scripture). Teach the scientific aspect of scripture with due diligence and let us "dumb masses" decide which side of the debate we'll reside on!!

Teach our children the *science* of *creation* vs. the "theory of evolution" properly, man! Remember this, you are supposed to "teach", not indoctrinate and train us with your thinking! Stop skewing the information to fit your agenda! I'll offer you another simple test, scientists.

Gather your pocket calculators, test tubes, Bunsen burners, beakers, telescopes, microscopes, laser scopes and satellites. Ratchet up the particle collider, the height of fearless arrogance, and ready yourselves. Gather your thoughts and gadgets and come with a ready prepared answer. Ok, are you ready? Here it is, so start calculating: prove God's Word wrong!

Prove that it is **not** better to give than to receive! Prove that doing unto others, as you would have done unto you, is **not** good! Or prove adultery to be good! Prove murder, lying, stealing, cheating and coveting to be good! Prove God's Word wrong, oh man!!! Prove scripture wrong or prove your theory right, for that matter, before you cram it down our unwilling throats as fact and truth!

Where is the evidence of God's Holy Word? Everywhere the **light** of creation shines! Where is the *proof* of your "big bang" **theory**? Nowhere to be found! Interesting! In the end you are only left with the ability to suspect, conjecture, speculate, hypothesize, theorize, acquiesce, capitulate and guess as to your notions of it all! It is time you stop training us to believe your theory as truth and scripture as myth. The fact is you can't prove or disprove either!

The theory of relativity, the big bang theory, string theory, super string theory, parallel dimension theory, quantum physics theory and the like are now properly labeled, **t h e o r y**!!! A "theory of relativity" is proof of nothing yet pride lifts you up doesn't it?

For me, when I read scripture I'm humbled as I wonder, and want explained to me, what is the *relativity of theory* when you possess the sword of Truth? I call religion and science, letters and numbers, words and equations equal building blocks in his mighty and majestic creation! And please understand that my definition of religion attempts to parallel that of God's, not mans!!

Unfortunately, as with science, man is equally guilty of tainting religion as well. Do you know why? Sure you know; it's the same reason as with science that I mentioned earlier, narrow mindedness: that label exclusively tagged only on believers, for some reason, as something to be avoided and shunned!

The fact is that we are all narrow-minded in our thinking so enough mud slinging! We are all guilty of being narrow minded and even hypocritical; another tag that, for some "unknown" reason, only applies to believers. But the truth is this; believers are narrow-minded about what they believe and non-believers are narrow-minded about what they believe. Fallen man truly is created equal-equally hypocritical and, in a bold thought process of needing to be right, equally narrow-minded! In retrospect I'd call you, Mr. Evolution, narrow-minded as well!

The bedrock of such labeling finds its' genesis in holier than thou attitudes; the paradox is when atheists juxtaposition reveals the label placed on people with holier than thou attitudes, bearing their guilt as well. They come across as made ill if the word Holy is even mentioned, with a condescending, holier than thou attitude. Why does this occur?

This occurs because man, in relentless fashion, holds one thing near and dear to his precious little heart. It spawns judgment, division and, yes, even narrow-mindedness! What is it? It is opinion.

We hold our own opinions in highest regard in every aspect of our lives! (With tongue in cheek I say) I can hear some thinking, "well that's your opinion", but it's true. One set of ears, one set of eyes and one thought processor, out of almost seven billion, is all we are each allotted.

Sure, we can meld, mix, mingle and agree with numerous other opinions, but at the end of the day, when we lay our heads on our pillows at night, we only have our own lone voices to contend with! We are all narrow-minded with only a one in seven billion ability to view life!

So we all better be sound in our thinking and opinions or others will sow seeds of rubbish! Think about it for a moment, our opinions are exactly what "the theory of evolution" desperately desires to shape, conform, train and manipulate into like thinking! Why else do you think it is so imperative to them to begin at such an early age? They seek to sow the seeds while the soil is still young and fertile!

However, with out your opinion, on any matter or issue, where would you be? And know this; our opinions are merely our internal summaries and views of God, life and man, in that order! Our "opinions" carry us, shape our thinking, guide our focus, mold our insights, direct our attention and assist us in evaluation and, sorry, even with all their benefits, cause us to be narrow minded. We all fall victim to the occasional eye booger.

So let's stop applying to *some* that which applies to *all*! Because whether you like it or not that is the fact of the matter! Now in light of this understanding just know this, our narrow minded, opinionated outlooks on life are just our way of conveying our very own "theory of everything", our own theoretical truth. And although this can act as a protective mechanism, when it overrides or omits altogether the Word of God, it has a blinding, lethal effect.

Scripture clarifies, with perfection, the proper measure to be used when dividing the Word of God. Believers, the next you time are labeled as narrow minded in your approach to God, life and man, find solace in what Christ declares in MATTHEW 7: 13-14. If you are anti-evolution, trust me, you will be tagged as narrow minded and bigoted!

Atheists, the next time you *label* someone *as* narrow minded you should read this same scripture as well; it is quite descriptive and specific about your future too! My point is this; we all bear the guilt of narrow thinking! What divides us is the focus of that narrow lens. Just remember this; "Small is the gate and narrow is the path that leads to everlasting life".

Now, before I move on to that "blinding science" topic, allow me to pour some fodder on something briefly. The crème de la crème of this ultra modern mental plane we now find ourselves in is one Stephen Hawking, Mr. Science himself, with ***his*** most highly touted and much publicized "theory of everything".

His "goal" is to prove that there is no God and no Wisdom involved in creating creation. And since the scientific community has propped him up as their fearless leader, they all bear the same narrow minded goal oriented guilt! However, their *hope of proving* God doesn't exist is shifting, I discern, into just a *hope* that God doesn't exist! A hope fueled by new discoveries, as with each new discovery they're beginning to realize the magnificent intelligent design to creation; what bothers me is their updated angle of attributing this design to E.T. instead of the God they hope doesn't exist. What a weird hope to cling to, as it takes mighty faith to believe and hope such things! It bears the characteristics of religion while denying and slandering its' purpose!

Now as far as I am able to discern, Hawk proves Al wrong and Al proves Isaac wrong and

Isaac proves Galileo wrong and Galileo proves, I don't know, Copernicus wrong and so on and so on. Each of these men was a great pioneer, but they've essentially been rendered as full of error.

Each great scientific mind that comes along, proves the last guy to be in error yet the scientific community holds these men up in highest regard, why? Hmmmmm!! I'll give you a hint; it's spoken of in the gospel and has to do with whether you seek praise from God or attention from man! But whom do **you** look to impress fellow comrade? Search it out for yourself and see. Just remember this; the **Word** teaches as the **world** trains.

Chapter 2: "Blience"

"Blinding science", thus "blience" is about as appropriate a description of this day and age as there is. We have all been blinded by the use, or mis-use, of this precious gift of science! We've been blinded, or more accurately blindsided by mans misguided use and *abuse* of God's supreme scientific architecture, Knowledge and Wisdom!

Everything that surrounds us blinds us and simultaneously gives testimony to the heights that man has ascended himself. Unfortunately our lofty heights are temporal at best. Now, let us begin to peel back the canvas and see, specifically, just how high this blientific fall will be. Let's examine in what ways man has stolen and called it his own invention.

And if you find yourself unwilling to part with your pride then stubbornly swallow that proud morsel down to your belly and accept whatever consequence it bestows upon you! Nevertheless, it is time Truth once again make her-self known.

When you lift just the corner of the canvas it starts to appear, can you see it? Can you hear it? Ahhh the sweet and savory anticipation for Truth! Let's begin with something near and dear to my heart, music. This is one topic you would figure doesn't even belong as a part of this "scientific" discussion but wait.

You see, music is also of and from the spirit realm, given to us, not something we invented on our own, as many suppose. Here I won't comment much on the human voice other than to state its' rightful position above all other "man made" instruments! There is simply no instrument fashioned by human hands that can claim to be its equal!

Its' characteristics, however, are cloned by those *man*ufactured when you factor in pitches, tones, scales and inflections. It's another gift from God that we've taken and called our own, manipulated and abused. I am a practicing musician; spending many a night practicing licks, scales, chords and the like on my guitar. Music's foundation is basic-a, b, c, d, e, f and g and variations of these seven notes.

I know personally how methodical, repetitious and structured music is. I am also highly aware of its' potent impact on our lives. It possesses the power to transform our very moods and thoughts! Do you not remember how David would play for Saul to chase away vexing spirits (1 SAMUEL 16: 14-23)? Spirit realms, good *and* evil, by the music we listen to, directly affect us! So wisely choose what you allow in, be selective! Need proof?

Scrutinize and pay close attention to scriptures' wording. After reading 1 SAMUEL 16: 14-23, grasp the spiritual significance scripture places on music, as it is a spiritual matter. You must understand this man; spirits, demons and fallen angels are constantly attempting to influence our thinking and music is the quickest way to achieve that end so be aware!

And how is music related to numbers? It's related through its' structure. In fact, that is what music really is, a structured arrangement of various notes, sounds, pitches, tones, octaves, flats, sharps, majors, minors, augments etc. It is, in fact, very scientific; do, re, me, fa, so, la, te. (I am only stressing the fact part to emphasize my scientifically based point).

All instruments: whether vocal, stringed, percussive, horned or keyed, are structured repetitions of a finite number of notes, scales, chords and beats. Very scientific and very abused! Remember, numbers are of God and are from the spiritual realms and thus find their origin there, not from us.

Sure, we can learn to play and teach others to do the same, but to claim it as our own? I think not!! That kind of thinking has only led to a "no holds barred" abuse of this gift bestowed upon us. We cannot claim ownership of a pre-designed, "built in", "hard wired" spiritual attribute!!

What we can lay claim to, however, is the heinous, blasphemous, wickedly vile lyrics we sing along with our music! Have you heard what the words of these songs typically spew? Have you pondered the meaning of the message they convey?

Individually they are wicked thoughts but collectively they brandish a vile, wicked, anti-Christ message! Contrast the song "if you seek Amy" with 1 JOHN 2: 22 followed up by 2 TIMOTHY 3: 2-5 for a fully detailed description of this generations' message filled fruit basket. To be sure, a decadent crescendo waits! Need more?

How about the song "Imagine"? You know it; "imagine there's no heaven, it's easy if you try, no hell below us, above us only sky"? This is absolutely anti-God evolutionistic thinking, filtered through radio and pumped directly in to the soul! What about songs declaring, "Jesus Christ deny your maker"? What do you make of someone who proudly announces they are "running with the devil"? And what of those determined to "shout at the devil"? All have overlooked a couple poignant little verses about slandering celestial beings (2 PETER 2: 10-14) and (JUDE 1: 8-16).

What do you make of someone who arrogantly declares they have "sympathy for the devil"? Or what of a song that boldly declares "I'm living for giving the devil his due"? This should not be a light topic of discussion! If those who slander celestial beings are allotted one form of warning, what will become of those who sing praises to them? Dangerous!

Scripture declares "Let no one deceive you with empty words, for because of such things God's wrath comes on those who are disobedient. Therefore do not be partners with them", (EPHESIANS 5: 6-7).

What of bands that name themselves **black** Sabbath or **Judas** priest or **Hell**oween or nine-inch nails, the size of the spikes used to impale Messiah to the stake? Like it or not; "**do not be partners with them**"!! They can all claim they are only "entertaining" the "dumb masses", but what they are really doing is spreading a sorcerised message of hatred toward God through the abuse and misuse of a God given talent, by way of scientific "invention"; all the while acquiring their fame and fortune.

To paraphrase: this is a prideful, lustful, boastful and greedy generation! There will be lovers of self, lovers of money given over to the lust of the flesh, not having that of normal affection towards each other but choosing the chords of sin and deceit. This is precisely as God's Word teaches would happen!

We sing discord in our songs, watch it through our TVs, indulge in it through our movies, train it in our schools, download it on our computers, and hence live it in our lives. Numbers and words hast we taken and defiled to our liking. We entertain our senses away! We entertain our morals away, all in the name of free expression and all by the means of our blientific "creations".

Has it ever occurred to anyone that sometimes it is better to not be heard or expressive? I'm not trying to fart all over your rights here; I'd just like to point out that God's Word teaches, emphatically, that "the tongue possesses the power of life and death"! And also declares, "By your words you will be acquitted and by your words you will be condemned"! According to scripture, going around expressing our opinions is more of a death trap than we care to know! *Gossip* anyone?

And for those who hide behind your "freedom of expression"; including but not limited to those in the music industry, you subsequently validate my point about our opinions! And you also confirm the notion of man desperately yearning to become a *god* himself, hence, the need to be *American* ***idol****ized*! But God is worthy of our worship where as you, o-modern-man, are not and never-ever will be!

What a quirky lot we are! Man doesn't want a God whose throne he has to grovel at the foot of, he wants a god who issues forth nothing but a never-ending flow of mercy and grace no matter what. Any mention of obedience or justice or wrath or accountability, on our behalf, and we're out the door!

But to have the chance to be on TV and grovel at the foot of the judge who can get him or her **idol**ized American style, yeah! They're in! Our desperate begging for attention is shameless. Where's your scruples mister and misses have to express myself?

Your desperate need to be heard and have everyone's attention grows from our deepest desire and need and wants to be worshipped and idolized and paid attention to! But your idol status is not merited, you've "evolved" into nothing more than a mouthpiece for satan himself!

And besides, the ability to write lyrics and sing songs is not your doing anyway so enough expressing! "When words are many, sin is not absent, but he who holds his tongue is wise" (PROVERBS 10:19)! Why does this not strike fear in the hearts of those in the "monkey business" of music? Because of evolutions' successful training program!

So of all the digitally enhanced, electronically mixed songs that the music industry produces I say this: if you refuse to praise God then bite your tongue and just play the “manmade” instruments! Enough “monkey business”! Spend time in the books of Wisdom: JOB, PSALMS, ECCLESIASTES, PROVERBS and the SONGS OF SOLOMON found within the Old Covenant before you spew your venom!

Until you create your own method of singing, and I mean the muscles and tendons and cartilage and blood cells that make up your vocal cords and gives them the ability to vibrate in such a way that produces music, the breath required to expand your diaphragm, *and create your own diaphragm for that matter*, then just keep quiet!

While you’re at it create your own breath too! Then create your own method of forming letters to words to sentences to songs! Until then, pipe down! Music I love when it is not stained with mans’ I, my, me (point of view=opinion) lyrics! If you will not repent of the vile words you contribute to mankind, than darkest darkness awaits you! And again, for us “dumb masses”, “**do not be partners with them**”; if you sing along you endorse the song! The structure of music is scientific; the lyrics, which accompany such structure, are the fruit that identifies evolutionary thinking via "blience". Here I will include a song I've written and produced that should convey the point I'm attempting to manifest. It's titled 'monkey man' in keeping with the monkey mentality we display, at the behest of the elite.

"The fruit of evolution, Monkey man….

'See no evil, hear no evil, speak no evil'; isn't that how the saying goes monkey man?

Hey there Mr. Monkey man

Your monkey voice speaks of how it all began

But monkeys can't talk so help me to understand

And what the heck is wrong with the thumb on your hand

You've been thoroughly trained to deny,

Evolution's left you mighty confused

Heaven's revolution's left you battered and bruised

You've been listening to the whispering lie

From a monkey to a *god* is the thought he brought into your head

He's still whispering in your ear 'hath God really said'

Hey there 'big bang' man of resolve

You say you came from monkeys then from monkeys evolved

But your link's gone missing leaving your puzzle unsolved

You're just a Creator hater as your theories revolve

You've been thoroughly trained to deny,

Evolution's left you mighty confused

Heaven's revolution's left you battered and bruised

You listen to the whispering lie

From a monkey to a *god* is the thought he brought into your head

He's still whispering in your ear 'hath God really said'

Mr. Monkey man your primate vision is blurred

He gave you ears to hear but you deny His Word

'Big bang' teachings as you speak the absurd

The cup of wrath is tilting cause it's done being stirred

God's about to reveal the lies

Evolution's gonna be exposed

And heaven's revolution's gonna come to a close

You still listen to the whispering lies

From a monkey to a *god* he's still hissing the whisper in your head

Still whispering eat the fruit, and be like God instead

MONKEY MAN!"

But that's enough about music; let's move on. How else does science effectively blind us?

Keep peeling and you find there are other scientifically enhanced fields that have blossomed and grown, not "evolved", in our "high-tech" world. What fields? They would be the entire entertainment industry, which includes all; sporting events, concerts, movies, television, radio, computers, cell phones and I-pods®. These fields of entertainment only draw our attention away from God.

Or how about the petroleum fueled transportation industry, which includes the shipping industry connecting: airlines with airports, trains with stations, and big rigs with warehouses and vessels with docks? And practical/leisure vehicles like cars, pick-ups, cycles and boats? Not to mention the big oil industries' fueling stations that keeps all manner of vehicles moving, in tandem with heating our homes.

Militarily with smart bombs, tanks, fighter jets, choppers, drones, ships, guns, equipment etc.? What about the restaurant and housing industries which include: the whole appliance sector, utilities, fixtures, building materials, food product and the factories that produces them, not to mention the tools to build them all?

The commercial real-estate industry which gives all products a place to be sold with all the: outlets, restaurants, strip malls, retail establishments, grocery stores, super centers and marts! And last but not least the banking and energy industries that fund and sustain it all!

As you can see all of these man made industries, born from sorcery's scientific knowledge, are intertwined and interwoven with each other, hence the height of our loftiness! I find it thoroughly refreshing what one little peel back reveals! It shows that man has, indeed, sailed to lofty heights!

Oh we have exalted ourselves much with our stolen, temporal ascension! Now would probably be a good time to take a look at and observe how high we've "ascended", through self evaluation not psycho analyzation, so we can discern just how this erosive scenario plays itself out.

What I want you to keep in mind fellow man is simply this: we are the created not the Creator! All that we have "created" in this technologically driven world is not real; it would all disappear in an instant if unplugged!

All, and I mean all, of the above-mentioned industries would utterly collapse! They all rely heavily on each other to function! What I mean is precisely this; there would be a cascading effect, which would essentially mitigate all other scientifically "created" industries!

I contend that the pillars for all of these are energy (electric and oil), again our "creation"! One field I didn't name is the medical profession, which could probably rival energy as a pillar especially when you factor in the subsequent food and drug industries with all their tentacles! Nevertheless we are masters at "inventing" things well enough but to *actually* **create** is God's and God's alone.

Just to make my point I'll ask some questions of the staunch, "Science is god", atheist evolutionists. When did your medium, whichever you choose, evolve in to being? Did a radio ever evolve into a television? Did a television ever evolve into being a computer? Or did a computer ever evolve into a supercomputer? I think you get my silly point, so I'll end with just one more question, albeit a difficult one to engage.

This is for you to give a full, honest and pertinent answer to Mr. "prove it" evolution atheist. Who is man that God is mindful of him? Read PSALMS 144: 3 then ponder, analyze, meditate, philosophize, hypothesize and theorize its' meaning before concluding you *evolved* the ability to do such pondering, analyzing, meditating or philosophizing!

Moving on, another tug at the peeling canvas reveals even more that's hidden. But instead of focusing on what *we've* "created", let's shift gears and look at created life itself as this will assist us in determining the difference between God's creations and mans scientific inventions. Where did DNA get its' structure? Where do the elements get their structure? How is gravity?

Where does structure get its structure? Where does water come from? When are the seasons scheduled? Where does photosynthesis get its marching orders? Who decided the degree of axis for earth and placed its' canopy of atmosphere around it? Why is gravity sustained? What designs the very breath you breathe? Where is your next one coming from?

And in that vein I'd like to impart some wisdom upon you. "Let everything that has breath praise the Lord" (PSALMS 150: 6). Do you recognize the implications of this Mr. evolution? It means with every breath you breathe you literally praise the sovereign supreme God I love and you love to loathe. Awesome! Even with all your vitriol and "theories" you still wind up praising our Creator!! Again I am going to point you to PSALMS 144: 3 and this time read through verse 4 as well, it is quite humbling.

And tell me this, how do you know if your heart will continue to beat one more day? How does the body know when, why and how to daily detoxify itself? Who placed our senses within us and why? Who programmed the human brain to accomplish all of its' multiple tasks? What purposes do memories and dreams serve? All of these are poignant questions that must be answered by those who claim to be in the "know"! Here's an idea, why don't you go ask a monkey?

And before I get a dissertation of how the systems all "function" together, ponder this; "In the beginning" **time** "God created the heavens and the earth" **space** and **matter** (thank you ppsimmons on YouTube©). There are some absolutely fascinating numerical attributes to GENESIS 1: 1 that are tied directly to the physical structuring of all that exists. There are many scientific "statistical impossibilities" found in this opening statement alone.

In agreement with scripture: time, space and matter are also the scientific boundaries in which everything exists as well. Now, with this in mind, if the "scientific community" cannot answer the above questions, which scripture does, than really, in what way does all this technology that they've "created" benefit us? The above measurably tangible evidence I've just asked questions about *predates* our ability to "create" anything!

Oh I've heard so many times how technology benefits us but I am curious to know how? Sure technology affords us many advancements but aren't they all just advancements in carnal comfort? Don't give a namby-pamby halfhearted answer to this question either!

Use your stringent mathematical equations, calculate it down to a common denominator and be honest, don't they only appeal to our carnal desires? This is not a benefit. They've exponentially rendered us lazy, selfish, and immoral and enslaved by "things"! Similar to Esau we have mortgaged our futures away for the temporal, which leaves us just as Esau was left: angry, defiant, outspoken, proud and hardhearted.

Through Darwin's theory we've waxed cold! We are now "stressed out", "depressed", "sleep deprived", "mal nourished", "attention disordered", "habit inflicted", "self-esteem deficient", "tired", "dis-ease riddled", "symptom infested" and "swimming in debt" because of our blinding scientific "creations"! At least according to this week's cycle of TV commercials and news (**N**orth-**E**ast-**W**est-**S**outh) stories. Again, just as scripture foretold thousands of years ago!!!

Now tell me, modern man, where did this modern era acquire *its'* genesis? How did mankind trudge along for thousands of years and then poof, morph in to a highly advanced, technologically advanced, "we're from monkeys" race? We did not; I repeat we did not evolve from nothing into something like the "big bang" theorists like to think is possible!

What man did was harness then manipulate; then manufacture electrical currents. From that day forward everything we live became possible. Now before delving in to this topic of discussion, which we will dig deeper into in the next chapter, let's preface something: electricity is also from the spiritual dimensions! Just like words, numbers and music, which we have also learned to harness and utilize and abuse, electricity does not find its' origins with man! It was bestowed upon us, man! And just like the aforementioned, man has tagged it as his own and abused it accordingly.

The point is this: within the last couple hundred years, this modern world was given birth and has grown exceedingly! This all began with the "discovery" of, "harnessing" of or more accurately put, induction of this spiritual "awakening" which powers us up. This is where "blinding science", thus modern "blience" began.

Just don't assume to know which side of the spiritual realms this "awakening" knowledge comes from! The Wisdom in the creation of electricity is God's and God's alone; the endowment of that knowledge upon man is not so easily detectable.

Unless you were there at the moment this event occurred, you can't know, but more on that later! For now, I suppose it would be the appropriate time to enlighten you in a certain matter; it is the matter of the tree of the knowledge of good and evil.

This tree can be easily broken down into understandable unity like this: your bible, scripture, the Word of God and all God's creation is the good knowledge. What is the evil knowledge? You're living in it, none other than man's creations; the manipulation of God's Word and creation.

Just look around at all this modern stuff we're surrounded by and find me one thing that in and of itself uplifts the glory of God? Name me one of mans "creations" or inventions whose sole purpose is to give glory to Yahweh. None!! All these inventions serve man not edify the body of Christ.

They are at our, as I stated earlier, beckon call, not for God's glory! This too should not be! It, technology, permeates throughout our whole being in an effectively subtle manner! This "tech" driven world we inhabit engulfs, in stealth fashion, every aspect of our lives.

You've heard the news reports that indicate a "return to the dark ages, if our electrical grid were to fail". This is a true statement people! We have been so lulled to sleep and hypnotized by all that surrounds us that we're unaware of its' prevalence!

We have fallen asleep! Wake up people! None of this is real, real in the sense that it is sustainable on its' own. Everything stems from our ability to manipulate and "plug it in".

Unplug it and we *are* back in the "dark ages"! Our heavenly Father's glorious creation, however, is absolutely sustainable in and of itself with no "plug" to be found, nor is our "belief" in it a prerequisite for its' existence, for it is He through TORAH physics who sustains it!

If all of our high-tech world were to be suddenly "unplugged" photosynthesis would continue. Do you know why? True science, conceived from the Wisdom of our Creator, will remain as long as Jesus deems necessary. All systems found in "nature" would continue to function as designed. Doesn't your mind comprehend this yet? Unplug our electrical grid and all systems fail yet within God's design it would remain all systems go. This is the dichotomy of God's science vs. mans' "blience".

What we, man, have "created" is hocus-pocus, magical, witchcraft smoke and mirrors with no substantial redeeming qualities at all! I do not; **do not** state this with a hope of it ever happening, I merely think it wise to be cognoscente of the fact. Nothing we've created is "real" so nothing we've created deserves our trust, that's all I'm saying! Trust God, not man, trust God's science not man's "blience"!

This brings me back to that poisonous monkey training of evolution, mans invention of imaginary proportions. Dissect the notion that you are from an "ape", therefore your life and all of life, for that sake, is no longer sacred. This thought process has led us down a path of complete moral destruction!

Life has no real inherent value; therefore abortion is all right! Life has no inherent value; therefore pack the elderly away. Life has no inherent value; therefore truth is subjective. Life has no inherent value; therefore marriage vows are overrated. Life has no inherent value; therefore be a human guinea pig for the agricultural and pharmaceutical industries. Life is void of any value, therefore, "live it up", "live life to the fullest, heck with the consequences", "throw caution to the wind".

"Go for it", be "lookin out for numero uno", "don't worry, be happy", "roll the dice", "speak your mind", "build up your self-esteem", "take the bull by the horns", "pull up the ole bootstraps", "reach for the sky", "shoot for the stars", "put your best foot forward", "step to the plate", "get in the zone", "all or nothing", and of course "just do it"! All of these "cliché's" are full of pride and void of God’s Word. They all inevitably lead to the attitude of “do what thou will”.

This is the place where all the “self helpers” and “gurus” dwell, and they’ll willingly share their blientifically-based knowledge with you-for a hefty price of course! They imply that you, oh man, are in control of your own destiny, a lie nurtured along by education programming us and training us to believe that we are monkey-made! If this were true then tell me, again, where will your next breath come from? You can’t!! Do you know why? Because we control nothing!! Control is an illusion.

This is YHWH's plan, his doing, his design, his guidelines, and his creation!! And no! The pottery can’t question the Potter. And before you arrogantly point thy finger and blame God for your woes understand this; every bit of this information and knowledge was stolen, not created, by man! We are the guilty party here!

We are the ones who have taken this knowledge and abused it to our hearts’ content! We are the fruit taker; that’s right we stole it! In that one single act of defiance by mankind in the garden, which includes you too modern evolutionist man, we trespassed the TORAH. It breaks down as such:

1st command-*we* trespassed it, thought we knew better thereby elevating ourselves higher than and knowing better than God. *We* are Adam, *man*!

2nd command-guilty, as our ***image*** of ourselves (self-esteem) was, and still is, higher than that of God!

3rd command- had a conversation *about* God (not with God which was normal up to that point). And we had this conversation about God, not with him, with none other than the serpent adversary, the father of lies! There was a **vain** mention of God instead of communion with him.

4th command-yep broke that one too by breaking his **rest**. The plan of salvation, the work of redemption was set in motion. Yes we broke and disturbed his rest!

5th command-we dis**honor**ed authority all together.

6th command-we **kill**ed man as a whole. For from that moment forward our penalty ultimately became death. Remember, we were cut off from the tree of life, hence death. Fallen man is dead to God without the redeeming blood of Christ.

7th command-adultery (idolatry). Yes, we tried to cheat on God and cheated all of mankind and their generations to come, out of our original inheritance.

8th command-we **stole** knowledge we were not meant to have.

9th command-Adam **lied** and blamed Eve. Eve lied and blamed the serpent. We lied about our accountability by placing blame elsewhere.

10th command-we **coveted** knowledge over Gods' Wisdom. We attained our knowledge while forfeiting his Wisdom with one chomp!

So I hope you can see we are not as innocent as we like to think we are! We turned our back on our maker, not the other way around, as some suppose!! But praise be to our heavenly Father for he loves us anyway!! That is precisely why he came to redeem us through the blood. Remember, man, "You were bought at a price" (1st CORINTHIANS 7: 23)! Don't think you are a *god*, man, for you are here because of him, not to be equal to, or usurp, him!!

I hope you understand my motivation for this tongue-lashing comes from a humble, genuinely concerned heart for my fellow man and for my family and me. We are all, <u>in fact</u>, in this together. I'll not pretend to know who is acquitted nor who is condemned, that is Gods' right, and his alone! I can only observe what is on public display since the inception of "evolutions" teachings and electric's "harnessing".

Look around at the world we live in and convince me there is no concern. It isn't good people! I'll float the idea that we are, contrary to popular belief, worse off now than ever before! You would think with all our high tech advancements that we would be far better off than currently in view, like science fiction depicts, but we're not.

I can hear the scientific community enlightening me that "you idiot,

science fiction isn't real"! I'll concede, but a spaceship on the moon was once science fiction, advanced weaponry was once science fiction, organ transplants were once science fiction, test-tube babies were once science fiction, electro-magnetic transportation was once science fiction!

In fact, science fiction is what mans' science strives to, and has always strived to, make real; make scientific reality, thus blinding science, hence "blience"! If we're honest about our reality, this entire modern world we've built up is the precipice of "science fiction"!

Sadly we are the epitome of a spiritually, morally, mentally, emotionally and physically bankrupt race. This is the Truth of the "creations", the "discoveries" and the "inventions" we've buried ourselves with, through blience! I for one find myself to be extremely concerned when I observe the identity eroding times we're in. And know this; observation is the key to it all! "Eyes to see and ears to hear", remember? So observe; the 1800's witnessed the introduction of three major shifts in our race: the harnessing of electric, the extraction of petroleum based products and the population explosion from 2 billion to nearly 7 billion people.

I'd like to end this chapter with a thought-provoking note: this is a cosmic war we are all entangled in. Once the already fallen cherub whispered, "Has God really said"-and we had the audacity to listen- we set in motion this whole yin-yang, universe of opposites, war of the cosmos. This war in the heavens is over us, man, and the TORAH and the GOSPEL are *for* us not against us. It is the war within the tree we chose, and sadly, if not aware and awake, still continue to choose.

Chapter 3: Full Blown Creation Manipulation

A long trip into moral decay down a path we dared to tip our toes on is where we presently find ourselves. As I scan the horizon I detect a disturbing by-product of three or four generations of identities being stolen, eroded and severely compromised. The erosion progressively began in the 1,800's and continues to bombard us today.

I'd like to drop a little hint about "identities" that we would all benefit from when perceived correctly. Know this fellow man; your identity may very well be linked back to Hebrew and Chaldean, thus Israeli, blood! That's right; you're not from an ape you're from an Abe, as in Abram! Yes the Hebrew patriarch Abraham, the "father of many nations" (GENESIS 17: 3-8)! And this would include the Chaldean matriarch Sarah, "the mother of nations" (GENESIS 17: 16).

We will expound more on this "identity" topic in chapter 9 but for now know this; knowing from whence we came helps identify who we are! And knowing who we are foreshadows where we're headed and since we've forgotten where we came from our future looks eroded and bleak! When we're born into blessing and purposely denounce that blessing and then turn and denounce the one who extended the blessing, the once blessed outcome becomes bleak!

Speaking of bleak, keen observation reveals it to actually be getting much worse with each passing day! And as days turn into weeks into months into years into decades, a startling picture emerges. It is a cultural picture of complete moral decay and utter bankruptcy. There is a societal break down of values and standards and it's been a long time coming.

Now to get an accurate description of the fruits we see, we must trace back to when this slippery slope was jumped on, and what do you know? That leads us right back to man's capture of a bolt coupled with his extraction of petroleum; That moment, followed later by the birth of evolutions' decrepit mindset, bound with the Freudian and Pavlovian psychological mindsets, has plummeted the population exploding human race expediently toward oblivion.

As you can probably tell, I harbor strong emotions on these particular topics: our eroding identities and our supposed "discovery" of electricity and fossil fuels. I realize I may not be very eloquent in conveying this message so I will set it forth in poetry. This is a little poem I jotted down a while back, which hopefully does my feelings on this matter a tad more justice! Master poetry it isn't, but it should present my "opinion" rather clearly, so here goes nothing:

The bolt kissed the "kite" on destiny's night

And the door of today opened wide

Manufactured light, much to our plight

Now there's nowhere to run or hide

Just plug it in and turn it on
Electricity readily supplied

Yet something's missing, something's gone
Somewhere man has lied

This charged world we live in seems so false
Convenience our God given “right”

But what you won’t find in this world that’s lost
Is Gods’ view of wrong and right

Spiritual encounters with the realm of mankind
When lightning arrived that night

The whole world shifted in a moment’s time
As our intellect of self took flight

To gain a full understanding as to where I am going with this just stick with me for a moment. I do not hold this moment of “harnessing” as either good or evil although I do have my suspicions; the truth is I don’t know, but it seems mysteriously shrouded and cloaked to me. I do know however, that it was not extrapolated on our own by a man flying a kite in a thunderstorm, unless that kite was in fact a celestially induced scientific experiment, by men with a certain “unsealed”, “last days” knowledge!

And a generation later the hidden knowledge was passed on. Ok! It was magically "summoned", but to claim to *know* the specific entity of origin would be utter folly on my behalf so I'll not succumb to such a temptation. However, I do estimate it to be ultimately of the forbidden tree in the garden as well! Then poof!* Nikola Tesla, a "clairvoyant", presented the ac/dc Tesla coil to the world and man's notion of reality changed forever. Can you say HAARP? Can you say earth resonance? Can you say frequencies? Can you say WI-FI?

I state emphatically that it is a spiritual journey we are still undertaking! Our electrically driven, wireless modern world was given birth over a century ago and the fruits are everywhere for everyone to see when one perceives it correctly. Bear in mind the decadent culture, which currently engulfs mankind, as we loftily purge upward in our descent in to moral decay, on our path of "enlightenment"! As I stated earlier people, it isn't good!

To really do my opinion justice I feel it necessary to just lay it out there for you as plain and simply as possible. This is an indictment on us all and I include myself every time I point the finger so don't think I hold myself in higher regard than everyone else!! I, too, am just as guilty of hypnotically gulping down this modern world we live in! I realize it's all we know and, yes, all we can know at this point in time so I am willing to concede that we are all on this path together.

However, I absolutely will not take ownership of the prevailing thought of wisdom about origins and apes and big bangs etc.! I am "in the world" but not "of the world"! For those who ascribe to this training and determine to train others to ascribe, you are on your own! Furthermore, this is a speck in all of our eyes, a pox on all of our houses, regardless of beliefs!

It can't be broken down into religious factions or denominations. Catholic, Protestant, Baptist, Judaism, Islam, Buddhism, atheist or Gnostic it doesn't matter! It can't be broken down into racial factions, economic status or geographical locations. It has affected and infected us all! You see, this modern age we reside in has its' genesis in the acquisition of indiscriminate electricity and the extrapolation and abuse of fossilized fuels.

It, electricity, is the source of power that keeps the whole engine running at hyper speed so I can't really discern whether it is bad or good, I only know it was prophesied! And its' counterpart, petroleum, is also a vital ingredient in keeping the engine running. Our energy industries make all other modern industries possible and as I gaze through the past I recognize the seeds of discontent being sprinkled down through time since their advent.

Here a little (evolution) there a little (remove scripture from schools). There a little (abortion) here a little (contraception). Here a little, sprinkle, sprinkle (gene splicing) there a little, sprinkle (remove the 10 commandments- under the guise of the "separation of church and state") and onwards and upwards. Harnessing our electrical friend and extracting our fossilized counterparts are moments of vast, life altering change, but the real corruption has taken place through mans' training along the way!

So I don't blame the pillars of oil or electric for our withered state, I just think it pertinent to understand where we are as a race since their acquisition. To get a better grasp of my point I have compiled a list of their impact on each of our lives. You decide for yourself whether this fits in to the general term "the theory of everything" which, by the way, would have to absolutely include the potential of *everything* to be possible, starting with the one who gave us the ability to theorize in the first place, God!

With each passing moment, we depend too heavily on "inventions" as we place our reliance in refined petroleum and electrical "output" and it would all disappear were we to be "unplugged". Our creation manipulation ability spells peril for our reliance on these two pillars of modern man. In no special order all:

Communication grids
Transportation grids
Banking system grids
Educational facilities grids
Energy facilities grids
Research facilities grids
Medical facilities grids
Nuclear facilities grids
Agricultural facilities grids
Water filtration facilities grids
Sewage facilities grids

Television station grids
Radio station grids
Computer network grids
Defense system grids
Fuel station grids
Telephone grids
Cell phone tower grids

We would lose every aspect of our modern existence, which would render us electronically helpless. I know generators would jump to the forefront but once the fuel to run those generators ran out so too would our ability to extract and refine that fuel. This would affect every one of us in differing degrees; some would excel, many would perish, many would survive but all would be affected regardless of religions or beliefs.

As you can see from this short list the losses would be heavy! It gets somewhat redundant but you get my point. And I am sure this list could be added to adnauseoum, for the reality is staggering when you think about it. I contend modern man relies too heavily on his modern inventions.

Through our pomp and arrogance we have surely elevated ourselves to lofty heights, all the while forgetting that it is not real! There is not one aspect of our being that would not be affected by a grand power outage! Is there a lesson in all of this? I think so, one that should point us back to God.

Again these are all things, and the real travesty is the training man has forged on us as he manipulates our very thinking, which leads us to believe we can't survive without these "things"! If that is true then how did man survive up to the point of these modern times? Wouldn't mankind have completely died off a long time ago if we truly lacked the ability to survive "pre modernism"?

Sure, it would be a stark, abrupt contrast in two opposing realities but un-survivable? I mean, we would have to literally "rough it" but boy would we be humbled once again. Maybe then we would turn our concerns toward seeking our true identities as pertaining to the coming kingdom instead of seeking only that which surrounds and serves us! Maybe that's what the powers that be have wanted all along. A dumbed-down monkey mentality society, so they can march on toward their fulfillment of the Georgia guide stones, is what they want?

Listen, I don't lack compassion for my fellow man and I don't mean to be insensitive, but how would anyone know Mr. Hawkings' opinion on any matter if the power were to go out? And how does anyone really know if that computerized voice box is really speaking his intended words? Are these questions too raw to ask? I do apologize and I mean no disrespect to Mr. Hawkings' condition, I'd just like to know since the scientific community hangs on this man's every synthetic word.

If it appears that I have a bone to pick I do! It's with those who doubt and question God and deem it necessary that all doubt and question God. "Misery loves company" for sure! If you, oh man, seek to doubt God and trust man go ahead and do so but for me scripture prevails for JOSHUA 24: 15 states "as for me and my house, we will serve the Lord"!

Circling back to a genre I "picked on" in the previous chapter, I would like to now call the rest of the entertainment business on the carpet for its' culpability in this downward identity eroding spiral we find ourselves in, which the kite kissing bolt makes possible; namely the motion picture industry and the television industry. But first the motion picture industry.

Movies! Ah the entertainment! Big score soundtracks, computer generated images, high tech graphics, spectacular special effects, heart wrenching drama, oh! The audio/video bliss we adore! We love it, we want it, we just need it and we're willing to pay hard earned money for it! Just look at how the home theater business has exploded over the last decade!

We dove in long ago and appear to now be taking on the characteristics of drowning in it or worse, losing what's left of this fourth, post electric generations' identity. Tink! We drown and aren't even aware of it!

With high-definition surround sound quality, movies make us laugh and they make us cry. They scare us, they pump us up and they bring us down. They make us think, they make us feel and unfortunately for many, movies offer "life lessons" which people take and integrate literally! Extrapolating even one "life lesson" from an in animate, nonreciprocal device or contraption seems to me to be a desperate attempt to pound a circular peg into a square hole. Tink, tink!

We don't even consider it odd that there are no more good guys vs. bad guys portrayed anymore, now it's bad guy vs. worse guy! It blurs the lines as our senses are dulled and "desensitized" away. I've heard it labeled as "the dumbing down of America".

I call it a silent subtle tragedy! Once the movies' "lines" were incorporated in to our daily language, which is altogether impossible minus electricity, we began verbally chipping away at our own God given identities. Clank! First, our identities as a race, human, are reduced to a trained monkey, metaphorically speaking! Second, our possible remnant status of the house of Israel, the ten lost tribes, is now many levels removed and forgotten!

We subtly and continually infuse and adopt "character traits" and "personalities" we witness displayed in front of us through the audio/video experience! How? We absorb through words and images. Remember, words are of the spirit realm so they place a spiritual weight of responsibility upon us as we use them! I will expound upon this verbal responsibility in a later chapter. Images and pictures are of the spirit realm as well, and so is our ability to fashion, create or replicate them. But the bitter binding takes place when we emulate, regurgitate and imitate them.

It is one thing to act out a scene from a movie but quite different when we go around habitually "speaking their language"! For instance: when Hollywood began blatantly abusing God's name, with complete contempt, vanity and defiance, society followed! Our audio/video bliss should not ever have this tight of a grip on our tongues.

Everybody gather around so I can whisper something in your ears, shhhh, and listen; stop heaping damnation upon us!!!!!!! People go around damning this in God's name; damning that in God's name, stop it! I even hear children spouting the same drek. Stop! You are not a monkey so stop acting like one, please!! Substituting "Jesus Christ" in place of a curse-riddled outburst is taking his name in vain!

It's simple, it's sad and it is absolutely true! Tink, clank, chip! The best indicator of the blasphemous success of evolution is right there where scripture teaches it is; the tongue! Part of

God's image stamped upon us all, evidenced through our ability to speak. You better read PSALMS 5: 9 before emulating what the Hollywooders speak!

To me it reeks of adversary, this "I'm a cosmic accident, nothing really matters" attitude, so "bite your tongue" society! And keep it locked firmly in your jaw if you lack the control to **not** imitate Hollywood! "Sticks and stones can break my bones but *words will never hurt me*" is an outright lie; words have the power to slay! "It is written"!

The images and sounds pumped out of movies are from the realm of what we are or were, I believe, never meant to take in and internalize! They're not real! They're images and sounds on a screen, given birth through someone's imagination people! So enough talking and acting like them, we have surrendered too much influence already! And as the movies have been detrimental, the next in line is downright lethal.

I'd like to turn our attention to yet another evidence of our compromised and eroded identities through electricity's strengthening. That would be the advent of the in home, wirelessly controlled, at our fingertips comfort of the tele! Our guilt in obeying this contraption is probably single handedly more responsible for our identity crisis than movies when it comes to morals, values, character and integrity. This mouthpiece of modern man has led us astray! Stop and read JAMES 3: 3-10 right now, all the while remembering that minus electricity they, movies and TV, don't exist.

The language, images and content of what gets pumped out of this thing man has "created" is abominable. It is the epitome and proof of the epistles being fulfilled, by way of its' programming standards being so low. The networks covet ratings not standards. They care not of your soul but of your "tuning" in capabilities. And don't forget the recent mandate for all TVs to be digital. There is a government mandate for television-that's just weird and highly suspect to me.

Like its' two industrialized brothers, music and movies, television is also given over to vulgar talk and display! The inherent danger with TV in contrast with movies is the daily intravenous drip television bestows upon us. The images are in opposition to modesty, the content and language in opposition to scripture! This opposition, coupled with the 15 or 30 or 45 or 60-second incremented undercurrent and underbelly of this "beast", is where we are ensnared with mind manipulation.

This is where the media playground of marketing and advertising resides via the TV commercial. For in commercials is where our mis-placed obedience factors in to the equation! Unfortunately we buy what it (TV) dictates we should buy! We dress how it dictates we should dress, we eat what and where it dictates we should eat and we entrust our hygienic cleanliness to whatever product it dictates!

We unwisely "overlook", as though blinded by the science of it; the pesticide laced, chemically enhanced, malnourished, artificially preserved, hydrogenised, artificially flavored, artificially sweetened, bleached, fluoride filled, pasteurized, genetically modified and dangerously poisonous side of the food and drinks we consume on a daily basis! Why? Mis-placed obedience to our created dictator!

We have collectively forfeited our own identities and replaced them with whatever this blasphemous "creation" of ours deems appropriate and necessary! Commercialism leads us around in little circles before a sharp left turn leads us right to the slaughter!

To make my point I'll look no further than the pharmaceutical commercials as they tout the medicinal "benefits" of the next generation of synthetically "created" substances! But have you ever listened to the "**possible side effects may include**" and "**may cause**" voiceovers toward the end of *every* one of these commercials? Scary! And it's with *every single* one!

The tragedy lies in the fact that people, we "dumb masses", obey and march right to the doctor's office and inquire of these medicines. Huh? Even though the damage of the "side effects" far outweighs the relief and comfort they promise, we line up as "sheep prepared for the slaughter"!

In fact, a recent visit to the doctor's office was an eyeball opening experience for me. I have always suspected we too easily lay our trust at the foot of the "white robed" among us but what I noticed this time was a big post office like, envelope "holding station" full of every pamphlet under the sun for every ailment under the sun! There is a pill for whatever ails you, man! And don't worry man, there's a pill for every successive side effect as well. Baahaaahaa-ooh ooh oooh!

I'd like us all to reflect on something for a moment. Our failing health breeds their thriving practice! And when you keep tracing back, our failing health also guarantees the pharmaceutical industry's success! So as TV pushes us in this direction remember this; in reality we are not a patient, we are a customer, and to them hopefully a repeat customer! Otherwise their office buildings would be empty. This means they callously hope for our health to fail, repeatedly! **Baahaahaa-**aah aah aah!

Not only is this hope securely founded, they can pretty much bank on it as they are keenly aware of our trained consumption habits. Our commercialized agricultural intake is loaded with disease causing agents and their pills bring a form of symptomatic relief to the ailment-laden masses. I hate to say it but; cancer is big business and diabetes is big business. So are osteoporosis and COPD and asthma and indigestion. Medicine distribution is big business! While the sheep suffer and eek out a baahaahaa, the "shepherds" hear "cha-ching"!

“Side effects may include, but are not limited to, four-hundred thousand things not at all related to the symptoms you bear”. “May cause a million other problems to arise, but that dastardly little symptom ailing you will subside”! We treat symptoms rather than causes, why? Our electrically driven television-istic dictator dictates such! Would we be at all surprised to realize one day that the electromagnetic currents and frequencies our modern world produces contribute heavily to all of our illnesses?

I’d like to share one more observation about television before I move on. Does anyone else find it odd, ironic, peculiar and highly suspect to have a news story about children being sexually molested when the segment fades to black, and is immediately followed by an impotency pill commercial? Again my brain goes, huh? That’s-but-should-what-huh? Odd!

It appears to me as though these commercial producers believe they’re advertising to a bunch of “dumb mass” monkeys or apes! Oh yeah, lest we forget, that is precisely what Darwin, Pavlov and Freud-based education has been methodically training us and manipulating us in to believing for decades! Oh the fruits evolution bears!!

Man, we better wake up now and take stock of our situation before it is too late! We better take inventory and evaluate whether what we allow to be pumped in to us, which could not happen on such a mass scale minus electricity, is of God or the adversary. And those really are the only two choices!

Whether talking about the food we eat or the liquids we drink, the prescriptions we ingest, the products we use, the chemicals we consume or the wretched content dumped in to us via entertainment, we had better begin to reflect on the dire situation we've "created"!

I find us to be guilty as charged, as found in Gods' Holy Word! All that we have "created" leads us in the opposite direction of our maker, leaving us in more peril than the three generations that have preceded us! So, is electricity the culprit or merely a conduit for such hi-jinx to occur? You be the judge of that. Tink!

I would like to repeat one last little tidbit of information before moving on. The advent of radio, satellites, television, computers, printing presses, movies, photography, flight, moon shots, automobiles, telephones and cellular phones are all relatively "recent inventions"!

Add to that; the medicinal chemicals, GMO's, aerosols, chemtrails, cleansing agents, hygienic products, agricultural refrigeration, nuclear weaponry and evolutionistic teachings all occurred within the last couple centuries.

And as far as we've been taught none occurred before the night the bolt kissed the kite that spawned our expedited degradation…in to a harvest of oblivion!

Chapter 4: Don't Be Us "Dumb Masses"

This is where the erosion of our identities should begin to really hit home and where we begin to see the whirlwind we reap from the wind we've sown to (ref. HOSEA 8: 7), as we "throw caution to the wind"! I'd like to share with you something I find utterly laughable, ironic and perplexing. It has to do with those who cry "no God in *school*, *government* or the *workplace*" or I'll sue! The same ones who want God gone are the first ones to bitterly scoff when Columbines and Va. Techs and Aurora Co. and Sandy hook's happen. The first to be baffled by an alarming rate of teenage pregnancies are the very ones who think it is good to pass out contraception to children. Tink!

The ones who hand their kids these video games to baby sit them, are the ones to shriek when violence visits their doorstep. They also allow their offspring to sit for hours on end playing video games, watching TV and afford open access to computers instead of getting outside to play, and then wonder why America's youth are defiant and overweight? Or is "morbidly obese" more soothing? They hand their children a cell phone at a tender age claiming "protection" as their motive, then are in a quandary when this device gets abused. Why? Why have we become like this? We bear a psychologically induced identity crisis.

Because coupled with big government is our mystified society, which has forgotten how to parent! And believe me when I say this; that's giving the benefit of the doubt to many who don't deserve it! We have two family incomes, which means someone's getting paid to raise our children for us, and we wonder why our children adopt characteristics contrary to moms and dads. Is it worth it? I certainly disagree with that notion. May I just notify your conscience real quick? We are not "dumb mass" primates!!

My wife and I live a more difficult life than most for this very reason, an unwillingness to sacrifice our children to Molech! Now I know I'm stepping on some toes!! Good! But like I said this "theory of everything" has to incorporate every conceivable outcome so take it in, digest it and hold it up to God's standard of living before you condemn. I'm merely attempting to jar us out of our comfort zone thinking processes we dwell in. As I said earlier, wake up, you've been lied to and "sold down the pike", fellow man! Wake up!!!

For us "dumb masses" this is a wake up call but for the politicians, judges, professors, lawyers, teachers, self-helpers, experts, media darlings, and agenda pushers (The powers that be, the "elite", the secret society members, the occultists, new agers, NWO Luciferian crowd) this is my response to it all!! To those of you who deem the possibility of finding your families' lineage within an ape's bloodline you sure do act accordingly! For those of you who don't aspire to this dreadful thinking, start acting accordingly! If God is your stronghold and fortress then start acting as such and proclaim it! Enough with the liberal theology! Enough with the trained monkey "syndrome"!

The atheist evolution-toting carnivores have had their day in heavens' court and been found lacking! To repeat what I declared earlier; I am not the judge on such matters, the Word is! And his Word teaches that those who deny the Christ already stand condemned, so save the rigmarole for judgment day. I would merely like to point out the fruits of their carnivorous labor.

If you've yet to figure out what I'm hinting at here I'll spell it out for you. The "elite-sovereign bloodline- powers that be" think they are superior to you and I. They see themselves as indispensable and you and I as readily dispensable. We are nothing more than monkeys occupying their space; breathing their air. That is why it is vital to shake loose the evolutionistic mindset.

What they've produced in us is quite startling as we now have: parents afraid to parent, too many who "spare the rod"! Teens who think everything is theirs for the taking, they possess an "I'm owed" attitude! We find children who think they *are* parents while some grandparents raise their children's' children and some grandparents are tucked away in retirement villages or old age homes.

"Common sense" is depleted! Confusion abounds and good is called evil as evil is called good. Ever heard that before? It's right out of scripture. Lest we forget so quickly what was reasoned previously about the garden, the good knowledge? Scripture is. And what is the evil knowledge? You're living in it.

Although it may feel like it; this isn't an attack on the "dumb masses", which includes me, although it is a wake up call. This is a post-dated preemptive strike, albeit 28 years late, on those in charge. Another group I have a bone to pick with is the "self help" gurus. You know them; they have life all "figured out" and know what's best for everyone. I know because I, regrettably, plucked from their tree in my late teens and early twenties, not knowing the damage inflicted by their trainings.

The dirty little secret is their *scientific* approach to life while hiding a form of spirituality woven in to their sales pitch! It's tricky, sneaky and it absolutely works! "Take action", "set goals", "no pain-no gain", "talk success", "dress for success", "be a winner", "can't never could", "reach for the stars", "reprogram yourself", "fake it till you make it", "as a man thinketh" and "think positive" are terms they enjoy throwing around, and for valid reasons. They work! But if you "fake it till you make it", when you "make it" you're a fake! There's no way around it, if you "fake it till you make it" you are a fake.

“Goal setting” works wonders for those who struggle to find purpose! “Taking action” is great for those who aren’t willing to wait on the Lord. Thinking “positive thoughts” is transforming for those who think, “Man is basically good” and only capable of doing bad sometimes. Enter that applicable word again, *man*ipulate.

All of these are cozy thoughts to harbor but they’re in opposition to Holy Scripture. And as far as “reprogramming yourself” why not start with abolishing all evolutionary notions previously programmed in to your thinking!! Pavlov, Freud nor Darwin are, nor ever were, superior to you and me so enough of the “blind leading the blind”, or we’ll all “fall in to a pit” (reference MATTHEW 15: 14)!

The modern psychological eastern mystacism forerunners like* L. Ron Hubbard (occultist) and Anthony Robbins (Hubbard endorser) are not concerned with your well being or eternal future. They are concerned with selling their products and ideas, *and* implementing them to children via education!

But God doesn’t require you to “set goals”, he wants you to understand and obey his will (ROMANS 12; 1-2)! God doesn't charge you to "take action"; his Word blesses those who "wait on the Lord" (PSALMS 27; 14). And thinking positive thoughts is difficult when you view man as God does. A good read of PSALMS 14; 1-3 should adequately notify your conscience of this truth!

A thorough reading of PSALMS and PROVERBS will bring conviction necessary to comprehend that man truly is "only evil all the time" (GENESIS 6: 5)! This appropriately presents us with our desperate need of redemption through the shed blood of the Messiah!

None of what the gurus and "experts" teach can redeem man! None of what they teach can remove the stain of sin because we do ***not*** possess the power to "reprogram" sin away! Their teaching can't save your eternal soul, a sober reading of MATTHEW 12: 43-45 outlines the dangerous outcome of self-reformation. The **Word** teaches, the **world** trains, remember?

Go ahead, do the "goal setting workshops" and psychoanalyze yourself but then ask yourself, will it be your goals or his will? Because that is precisely how it lines up, your goals vs. God's will! Oh, they didn't tell you that did they? Scripture does!

All of these catch phrases can be the trigger that helps us manipulate our emotions and, sure, they make us *feel* better but they fail in the task of salvation! They lift up our pride and self-esteem yet fail in either bringing glory to God, edifying the body of Christ or offering man solutions to his desperate need for salvation! They may sound good but they are designed for the carnal mind, not the heart or spirit!

Again, for those of you effectively trained and manipulated, those who think your DNA can be traced back to a monkey, have at her, but for those who find their faith firmly based in scripture keep this in mind; all of their guru techniques are influencing the educational systems we send our children to every single day!

I do not advocate rebellion or upheaval; I just think we should be aware of what is going on in our school districts. They have absolutely adopted some of the "self help" methodology. Do you know why? As far as learning skills go, they work! However, learning isn't the issue here, what's being taught to our children is the issue! Indoctrination through repetitious training is the issue; don't lose sight of that!!

They continue to train them to believe that they are from apes, not made in the image of our Creator El Shaddi! They train them to believe that their self-esteem is of higher value than what is right and wrong. They train them to believe that *safe* sex is just better than abstinence and they train them to think that you, parent, become more irrelevant with each passing day! No wonder we trail the world in edumication and lead it in illiteracy, unwanted pregnancies, obesity, suicide, violence, alcohol consumption and substance abuse. There is too much training and indoctrinating, and not enough teaching going on! Teach these children the methods of how to learn not the methodology of what to learn! Teach; don't train!

They train children to believe abortion, shrouded in "women's' rights", as okay and even kept secret if need be. They train that daddy and daddy at home or mommy and mommy at home is normal. The progressive, kum-by-ah, can't we all get along, yuppie spawning, peace for all Woodstock generation, politically correct crowd, really strikes my ire!

I apologize for going *almost* political momentarily but I have to settle this out. Do you know both *eliminating* God from school and *inserting* evolution in school were doled out from a courtroom bench in the 60's? They were not passed into law by congress, or the senate. The court system enacted these laws on behalf of someone as the House of representatives was completely circumvented and legislation was by-passed.

In the same vein of courtroom quirkiness, has anyone else observed the courtroom ritual of scripture opposition? We, believers and non-believers alike, are to place our hand on the bible and swear to "tell the truth, the whole truth and nothing but the truth, so help me god". Yet open the sweared upon bible to MATTHEW 5: 34-37 and it declares we ought not to swear at all. God's Word and America's courtrooms are not very good siblings. They both contain law but one is life while the other is subject to man's action upon it. And on behalf of evolutionist thinking, the Ten Commands aren't allowed public display in many a state's courthouse.

Since our government chooses to progressively ascribe to the evolutionists agenda, they had better take a good hard look at PSALM 14 themselves in its' entirety! These suicidal, detrimental acts were not voted into law. They were forced upon us in 1963 (Abington vs. Schempp-prayer eliminated) and in 1968 (Epperson vs. Arkansas-evolution instituted) by a conglomerate of progressive judges in agreement with the atheistic agenda, and some how became the law of the land.

On a side note I'd like to comment on "women's rights" briefly. I think it has failed miserably! As with most things in life, the idea started out well enough, but look at where it has led us! Woman, in her quest to be considered equal *to* man, instead became equal *with* man.

What fruit has this wrought? Well, woman is shamefully just as much a pig as man now! I do apologize if this seems a little too intrusive but it remains true nonetheless. Sadly woman, in keeping with atheist evolution thinking, is as far off the path as man now! When women's rights became women's "lib", the bottom began to fall out!

I am not at all, in any way, against women's rights as long as it is bound and adhered to according to scripture. It's women's "lib" that veers off the path. Fellow woman, I do not hold ill will toward you whether you are the president of a country, president of a corporation or a stay at home mom as each are a woman's right, by the blessing of God.

However, those who progressively prop their "lib" up have left the bounds of scripture and now man and woman are equally narrow minded, equally identity eroded and equally as far off the mark as each other. So congratulations woman and welcome to the time honored man club of pigishness, I say sarcastically.

Lastly on this issue is the cloaked insertion of abortion, 1973 (Roe vs. Wade) within the boundaries of "women's rights"! These two issues are *not* one and the same, as the agenda setters would have us believe! Life's right to exist supersedes womans' right to make up her mind on an issue every time! And before tagging me as a male chauvinist, I hold the same true for men too! Missions are aborted, not lives! Come on people wake up!! And besides, Jane "Roe" is an alias anyway so the whole premise is deplorable.

And life's' right to exist can only be impeded through murder! So the next time you hear a debate coined as "freedom of choice" vs. "right to life" seek God's "opinion" on the matter! On top of that, the Supreme Court was never given the power to *enact* laws protecting "freedoms" or "rights", they're congressional and senatorial issues! Judges must *uphold* the laws not rewrite them! Oh the 1960's progressive, liberal theology movement! Talk about itchy ears!

Now back to the two mommies, two-daddy syndrome. This is another issue being played out in our court systems! Mommy^2, daddy^2? Huh? What's next? God bless Gomorrahica? Same sex marriage? Really? With all due respect, if the homosexual agenda really gets its' way, what would be the end result? A race that refuses to replenish itself will inevitably cease to be a race anymore (Of course that's where the elite god complex scientists could regenerate their perfect, AI, bio-engineered, super-human chimera race), at least until the singularity is reached.

As the evolution mindset flourishes confusion abounds! And one area under constant assault is the sanctified structure of marriage. You see there is a huge misconception in our culture when it comes to honoring our marriage vows (the pinnacle of man/woman equality), which is the foundation of any good family life.

This, for me, excludes our present "need" to "redefine" marriage, which somehow only recently became necessary in our country. An in-depth reading of LEVITICUS 18 also *excludes* the *need* to "redefine" marriage! I'm perfectly content, as we should all be, with God's boundaries of marriage. And for those of you on the opposite side of this issue know this: I do not care how you feel, I will not be politically correct-because I'm not a politician-and I will not compromise my affinity for God's laws and precepts to fit your agenda!

Now that I'm done being *almost* political let's oppose another one of evolutions' most successful areas of attack, obedient spouses. This also includes the ultimate success rate of the "self help" marriage counselor "expertise" vestibules with their 50% marriage failure rate! Confused?

For the marriage counselors, experts and therapists to have a *successfully* thriving practice society's marriages must be "on the brink", "nearly in shambles", close to *failure* and on the path to divorce. Still confused? An expert or counselor can't fix a vow that we've spoken then broken, as it is a spiritual matter! As patients are to doctors, embittered spouses are to them, merely a customer! Anyway,

Wives, you are charged in scripture to, shhrrieeekkk, "submit" to your husbands. But for some uncanny reason the verses that follow this submission are always omitted. I think this is mostly mans (male) doing at countless weddings, on countless occasions throughout our land.

Wives are to submit to their husbands, but here's the part we never hear, and the husband is to love his wife as Christ loved the church, as laid out in EPHESIANS 5: 22-28. Changes the whole context doesn't it? Men, if you refuse to subject yourself to Christ's authority then you forfeit the subjecting of your wives to your authority, period! This might surprise you as I jog your memory husbands, but Christ died for the church, the body of true believers.

You can seek the services of any "expert" marriage therapist or guru you want but if you ignore scriptures' outline of marriage, it won't succeed. A good reading of 1st CORINTHIANS 7: 1-6 on top of EPHESIANS 5: 22-33 will swiftly put this all in to its' proper perspective. While you are at it read COLOSSIANS 3: 18-19.

As I said, we have a fifty percent divorce rate because we choose to follow the trainings of man rather than the teaching of our Creator! Who knows the better way? We have skyrocketing pregnancy rates among young, unwed girls because of ignoring God's Word. The most detrimental part is the success of re-labeling this particular sin as a "mistake"!

We have slaughtered countless numbers of unborn children due to our propensity to just *not* heed God's warnings, all disguised under the banner of "women's' rights"! Oh the heights man (which still includes woman) has elevated himself unto in his never-ending quest to be a *god*.

From our elevated violence, elevated wickedness and elevated corruption, with elevated alcoholism and elevated drug abuse leading to elevated *oops!* Pregnancies, elevated std's and elevated broken marriages culminating with elevated strife, elevated child abuse and elevated broken homes, our elevated esoteric pride, and ultimately elevated folly should deem the term "reap what is sown" as pure Truth!

All of this elevation leads to; elevated stress, elevated blood pressure, elevated worry, elevated confusion, elevated doubt and elevated doctor visits, elevated medical conditions and so forth and so on. And most of all it only elevates the height from which we will fall! We've all *fallen* victim to the theory of evolution but hey, when you start to believe "monkey see-monkey do" what else can you expect?

There are two other tragic by-products of this digital age that threatens the last bastion of our identity and no it's not "global warming"! It has to do with the next generation of agenda setters, none other than our teen-age populous. Of course this does not include every single teen, just a vast majority of the, text message while driving gang!

And if you scan just beyond the horizon at the ones who are now yet babes themselves, it's even scarier! These are the little ones who spend hours on end camped in front of the video game enterprise with its' maniacal output of disturbing content. But I digress; let's focus on the next generation of partakers in the "monkey business", i.e. teens.

Not only do they *not* ascribe to the standards of generations past, they absolutely think, believe and demand that their desires be met…. Now! Right now!! They aren't the least bit aware of the fact that there has even been a moral degradation of our characters, values, standards, society and identities! Clunk! Chip! Tink! Clank! Manipulated and trained through and through! Check mate?

Are they perhaps the ones to finish it off and seal the deal? Teens and their palm held cell phone/computers are inseparable! They (cell phones) are their "life lines" and we their parents-with our God-given position of authority, are expendable, which is a prophecy foretold some two hundred decades ago! And bear in mind the simple reason they all wield their palm held contraptions-television dictates it!

With each push of the proverbial envelope in to the abyss that the digital world inundates us with, we rapidly disappear! Chip! Teens have their computer "fix" housed right there in their cell phone computers now! It is always with them, always on and it has become *their* deity!

They are a frighteningly violent bunch and look to harm people for amusement. In a recent newspaper article my wife shared with me there is an article about 11 and 12 year-old children who beat people for fun and games. Apparently it's called "catch and wreck" or "catchin' rep". The article points out these are 11 and 12 year-old boys *and girls* seeking violence as a means of fun. We truly are equal now aren't we women? Scripture tells us; we reap what we sow!

There were a 42 year-old woman and a 73 year-old men who were attacked by these kids, for fun! The 73 year-old man is now on a ventilator because of this attack. No scruples, no fear, no remorse, no concept of right and wrong! This is the inevitable fruit of a "spare the rod", psychobabble, "self esteem" building, evolution drenched and liberally theologized mindset! This is the pedigree of their training.

This is both frightening and foretold in scripture. They have life all figured out by twelve and bear no shame in "expressing" it, because this is also precisely when their inexperienced opinions take flight. They want the power of God-given authority; yet want no parts of, nor are they capable of, the responsibility that comes along with it. And unfortunately for everybody, they are the primary targets for the whole "monkey biz"; yes the "show biz" industry!

Quite frankly, I'm tired of the multitude of parents who are enablers for their **kids**! Where are the men and women with a spine? Instead of holding these kids to a standard, too many parents lower or eliminate any standards altogether! I'm telling you, scripture declares otherwise! Parents; you are not your offspring's' friend you are their parent! There is a severe penalty on its' way for this evolutionistic, psychology laden thinking! Again, WE REAP WHAT WE SOW!

Teenagers speak, if text-messaging can really be considered speaking, an acronym language hidden and coded with letters, numbers and symbols crisscrossing, re-defining, reversing and mutating before our very eyes. It is a fluid situation in the theater of battle for the souls of our children, and it's a direct descendant of our psychologically marinated minds, shaken *and* stirred with atheistic evolution training! This "spare the rod" mentality is a major contributor to why our future looks so whirlwind-ishly bleak, as alluded to in the last chapter.

"Thus, by their fruit you will recognize them" (MATTHEW 7: 20). Just listen to their speech-*tragically parallel to the babble in entertainment*, observe their attitudes-*ripe with evolutionist jargon,* see and hear their plight-*flooded with psychological excuses like attention deprivation* and realize that they will progressively contribute more and more with each passing moment to the elevated statistics mentioned earlier! Again, read MATTHEW 7: 16-20 for fruit identification.

They bear the fruits of iniquity, anger, strife, violence, depravity, lewdness, selfishness, pleasure, fearlessness and knowledge. We need to get off this path of "monkey see-monkey do", eliminate all "monkey business", drop to our knees in prayer and grovel at the foot of the stake/tree/cross for restoration! And we begin this by putting wicked idolatrous evolution away and picking up the Word of God again and learn real science from his Holy Word, the sword!

The unwillingness to do this, adhere to scripture, is one of the results of drinking the potent concoction of Freud's, Pavlov's and Darwin's teachings, mixed and mingled and manipulated and spoon-fed, via education! This is the melding of Pavlov's discovery and Darwin's theory and Freud's slip and, unfortunately, man has allowed himself to become the intoxicated object of that experiment.

The sub-con**science** and con**science** mind are manipulate-able, period. Effective evolution babble trains us to believe that we are merely evolved animals, contrary to what God's Word declares! In essence, we've allowed ourselves to become psychological Velcro®; theory inundated test subjects on a nation wide scale.

Couple that with the notion that we are therefore "trainable" and "programmable" and presto, we have the suspension of our beliefs, and in natural progression the degradation of our faith! We have at the pinnacle of mans' curious nature, psychology. And when you slather on top of that the media blitz of entertainment, we're inundated!

Why is any of this pertinent? Understandings; when we understand what is really going on, and conclude that we are, in fact, **not** "dumb masses" we're more prone to better decision-making, at-least that's my hope. Listen, all the psycho jargon babble we adhere to is their condescending attempts to convince us that we really are nothing more than animals! It is identity theft on a grand scale.

The self-help lingo, which is just an updated, advanced version of this psychological mind-set, deals only in our mental faculties. It's so effective because it doesn't dare to offer solutions to our condemned hearts! But our hearts is where the problem arises, "For out of the heart come evil thoughts, murder, adultery, sexual immorality, theft, false testimony, slander" (MATTHEW 15: 19).

Sure, we can go around all day long "feeling" good and even very easily convince ourselves that as long as we "feel" ok everything will be fine but that's just not true! Life is not about what they've made it about: **success**, settling down, **mortgaging** a house, **financing** a vehicle or two, and starting a family and such. Yes, there are elements of truth to every inclination we have, but absolute Truth is where our Creator lives and he wants us there as well!

Now before you go off concluding I am an anti-family man know this; I'm happily married with six children! I absolutely adore and believe in family. But we're on the path of having to redefine marriage now, remember? That is an argument I will not even consider entering! God made the bounds of Holy matrimony man and woman, not what presently devours our resources and courtrooms. I'll not be manipulated in such a way! Why? I'm not from a chimp, gorilla, orangutan or a monkey; I was created higher than them, got it? And so were you!

Here's a zany little question for you, man. Why, when discussing God and religion, is man so bent on his right to the *free will* God gave him in the garden, yet tell man to not be an adulterer, a slanderer, a thief, a liar, a homosexual, a drunkard or drug abuser or impulsive or compulsive and his lack of *will power* is excused away as a disorder or an addiction? Why? "Sin is crouching at your door; it desires to have you, but you must master it" (GENESIS 4: 7). It's much more soothing to claim our vices are out of our control than it is to own up to our vices.

Just let the psychological, analytical self-help gurus' **training** go in one ear and out the other; and **teach** your children to do the same! Do you want to raise your child to be the head of their class or raise your child to everlasting salvation? There has been enough consumption of their maniacal, deceptive, manipulated and scripture scourging trainings!!

We must eliminate the mind manipulation and training taking place and return to the foot of the one whose image we are made in! I realize I've used the terms manipulate, manipulation and manipulated repeatedly, but by the end of reading this book you should have a good understanding of why it's vital to comprehend its' impact! But for now breathe for there is a reprieve.

Oooh oooh
oooh aaah aaah
aaah?

Chapter 5: Graceful Equations

Ah the chapter of grace, the number five representing graces. Now is where you can begin to see the real science of the bible emerge! You'll see how God has applied numbers to his entire creation. Here you'll perceive the antithesis of what we've been "taught" in schools and universities across our great land.

This is what you and I should have learned decades ago had science and scripture been properly taught, side-by-side, instead of opposed to each other!! This is the point where we drop "opinions" and pick up the raw facts. And this is the point of no return for me, as I realize this will be foreign information to most churchgoers, as well.

Let's enter the real realm of numbers, which, oddly enough begins with language and vice versa. As stated earlier these two, numbers and letters, are intertwined in a marriage of perfect harmony in God's perfect Word. How? Open your eyes, man, for here comes Truth; Hebrew.

The Hebrew language is the pinnacle of them all! It surely is a mysterious, spiritual, multifaceted and structured language like no other: the most similar as far as letter-number equivalence being the Greek language. I've also read it's been estimated that up to #75-80% of our English tongue can be traced to this ancient Hebrew text, which lends viable credence to our "Lost Tribes of Israel" status! So I'm not going to haggle with the "prove it" pioneers on this one, study it for yourself if you doubt!

What I will do is point you in the direction of real "higher learning" and it begins with the Hebrew language and alphabet! This truly is an amazing structure of letters, definitions, meanings, words and values! It is the foundation for all of God's creation! Not to mention that it is more than likely our native tongue and dialect, since upwards of ¾ of the English language is derived from it!

From the Aleph (Beginning) to the Tav (End), yes, as Christ taught, "the First and the Last" letters of the Hebrew alphabet! The exclamation "I am the Alpha and the Omega", as translated in Greek, has a much deeper truth than we normally take note of, but Messiah said it so let's discern its' hidden truth.

Of a truth I tell you, the Aleph to Tav, or its' Greek translation the Alpha and Omega, is found in the very first verse, and I would say all encompassing verse, in the bible. But you can only literally see it in Hebrew. Now, I am in no way claiming to be a Hebrew scholar. But I yearn for you to see the seal of authority and authenticity of the Holy writ.

Reading right to left it states "**תאו ץראה תישארב ארב םיהלא תא םימשה**" ("bereshit barah elohim hashaymim wuet haeretz"). "In the beginning created God the heavens and the earth". Again, I am no scholar, and I won't haggle with waw or vav pronunciations, or transliterations. I want you to focus on the two untranslated letters smack dab in the middle, the exact center of this mathematical structure, and you find the two letters **תא**, the first and last letters of the Hebrew alphabet-Alephbeit. They are the letters Aleph and Tav , the first and the last, the beginning and the end of this Hebrew tongue.

Yahshua Jesus the Christ, right from the start, seals this book we call the Living Word of Yahweh God, but you miss it completely in our English translations. "I am the Alpha and the

Omega (the Aleph and Tav), the First and the Last, the Beginning and the End (REVELATION 22: 13). This is sealed from GENESIS to REVELATION by its' Author.

After his resurrection, Yahshua Jesus spoke these words to define himself, why? Well, the Hebrew language was the language God chose to dictate his Word to the Hebrew people after leading them out of bondage and through the wilderness and to the Promised Land. These are the letters, words and the language our Creator chose.

And what happened centuries later? Precisely as his Word prophesied, the Word became flesh! The Aleph and Tav, Alpha and Omega, the Beginning and End of every matter became flesh and walked amongst us, hence, Immanuel (God with us)! That's why Christ defined himself this way after his resurrection and ascension! The work of redemption was completed from Beginning to End, as the one to come came and cried out from the cross-, "it is finished", fulfilling the spring feasts of the LORD.

Do you discern the connection on a deeper level yet? To grasp this profound Truth one must begin to utilize the "language" of numbers. From Aleph to Tav the whole message of the Word was completed.

The literal Creator of the literal universe literally walked as one of us as foretold in his Word, his Hebrew Word! The Old Covenant, written in Hebrew, prophesied this Immanuel would one-day step on to the scene to accomplish Aleph to Tav, the very Word of God! The one to exclaim "I make known the end from the beginning, from ancient times, what is still to come" (ISAIAH 46: 10), arrived and fulfilled it!! So let us begin to examine the Hebrew dialect.

This peculiar language consists of a twenty-two lettered alphabet beginning with the letter Aleph and culminating with the letter Tav. What we were not taught, which the writers of scripture absolutely had knowledge of, is that every letter has multiple meanings.

Each letter has built within its' structure: corresponding number values, place values, definitions and meanings. Each individual letter has multiple dimensions to it. God's Word has multiple dimensions to it! This book we possess called scripture has a multi-layered, multi-dimensional, single-purposed message to us. It is his message to us in multiple depths, all housed in this one book we call the bible. A **cautious** read of sefer yetzirah, while diligently discerning its' origin, can assist you in getting into the frame of mind necessary to understand the depth of this language. I do not adhere to Kabbalistic teaching, as they deny the Messiah, but I do agree that their take on the 22 letters of creation ought not to be swept under the rug; especially in light of Christ defining himself by these 22 creation letters.

Here is another magnificent and mysterious, yet enthralling, structure to the Hebrew Scriptures; PSALMS 118 is smack dab in the middle of the bible but chapter 119 is the beginning of something altogether different in that it acrostically encompasses the entirety of the Hebrew alphabet-Aleph to Tav-in this one chapter. This is profound and speaks to the divine authorship of the bible! It is also profound in revealing David's heart for God and his Word, a heart all professing believers should possess!

Before you dig deeper into this jaw dropping, amazing, thought provoking information please allow me to build a bridge of continuity with two "schools of thought" so prevalent within the church. I'd like to dispel an argument that presently tears down instead of edify the church. Do we read scripture "literally" or "figuratively"? That's like asking God "do you want me to take you at your word, God"? Unless it's prefaced otherwise, as in JOHN 10: 1-6, 16: 25-29 and GALATIANS 4: 24, take him at his Word literally!

Do not forget how Christ *literally* responded to the adversary's temptations with "it is written"! The answer to this question must always be yes! Take him at his Word, for "it is written"! Seems like a silly topic to even ponder so set the record straight in your self right now!

Stop attempting to interpret scripture erroneously and begin to divide the Word correctly. Start from the frame of reference of always literal and divide from there. The Word does not twist from literal to figurative, as many assume, it's always *literal* and then moves back and forth between literally *spiritual* and literally *physical*.

Clearing up this one error can potentially open up multiple levels of the scriptures' Truth. Do not forget Christ's attributes as 100% God and 100% man, all God and all man thus, *literally*, all spirit and all physical. "The Word" spiritual "became flesh" physical; so when reading scripture please take him at his Word and remove this division from within the body. God is not like man, when he speaks he means it literally for "Every Word of God is flawless" (PROVERBS 30: 5). The double-edged sword cuts both ways and is applicable in both spirit and flesh.

Is scripture sometimes allegorical? Absolutely. Are there idioms in scripture? Yes. Are there figures of speech in scripture? Yes there are. Is there symbolism and symbology incorporated in the text? Yes. Do any of these ever over-ride the necessity of taking God at his Word? Never.

With no more delay let us exit the modern realm of theories, namely evolution-which wants desperately to separate God's works from God's Word, and enter the realm and mindset of the writers of scripture. Yes, put down the milk and pick up the meat.

Bible codes, gematria, "theomatics", "shemology" (name-ology-*only as applied in scripture*), numerology (*as it applies in scripture, not fortune telling*), astrology (*again only as applied in scripture*) and even "geological" examples of scripture are vital in re-melding his works with his Word. And it is essential in beginning to wash away the blemish of evolution! Now I know right away these will scare some away but please research it first, we are talking about eternity here.

On a semi-side note here, I feel it necessary to briefly address the topic of astrology. Did you know that as a Christian you are to be your own astrologer? The sun, moon and stars are to be watched, observed and retained by Christians; not worshipped. It's all a cosmic clock that informs the believer to prophetic events so be aware.

These are the scientific aspects of scripture satan, via modern man does not want you to learn. Understand this though; the writers of scripture possessed this knowledge as they wrote the Torah, the historical books, the prophets, and the books of Wisdom, the gospels, the epistles and even REVELATION. This knowledge was known by them so don't be spooked, learn!

As to do these topics justice I'll put my opinion away and simply point you in the direction of the material you can acquire at your own leisure to learn more. But many in today's' society claim the bible, and especially the Hebrew dialect, to be "outdated" and "not applicable" to modern man without ever researching it themselves. Therefore they take the above topics and abuse them outside the confines of scripture, which is extremely dangerous!!

They seek to know the future, to tell their fortunes and grovel at the altar of their daily horoscopes! Do not do this!! With due diligence do the homework for yourselves properly, and see if God's Word is "outdated". You will be very surprised at the results and hopefully begin to understand and comprehend and discern the different levels present in God's Word. This is not for new believers to be "dabbling in" for it is a deep spiritual matter of confirmation! This information is not intended to assist one in drawing people to Christ, as that is Holy Spirit's doing. This knowledge is conformation to strengthen mature believers.

Under no circumstance do I advocate this to be done haphazardly as it is much too dangerous! Actually, I would advise you earnestly pray for Holy Spirit guidance before beginning this endeavor. I wouldn't even step on this path if you sense God's Holy Spirit prompting you not to, but make sure you "test the spirit" before concluding either way. For believers I plead with you to proceed with caution, but for the atheist evolutionists I ask you to purposefully dive in, "Throw caution to the wind" and "go for it".

My stipulation for both groups is this; you must, I repeat-you must utilize scripture as your foundation! Believers, listen to the Holy Spirit guidance of El Shaddi! The only way to do this is through much scripture reading and prayer. In fact, I would advise a complete read through of the entire bible at least once before beginning such a study.

Verify what you will be internalizing with scripture and, again, test the spirit! The Word is the hub and the reference materials are the spokes, not the other way around! If you approach this backwards you risk irreparable damage! Don't grieve the Holy Spirit by gambling or rolling the dice! Fearfully and humbly approach the mercy seat to learn! Reference materials?

Books. Books? In this HDTV, surround sound, computer age you want me to read more books? Yes! There are numerous authors who have written extensive literature on this hidden topic of God's scientific presence everywhere, which are readily available:

Ivan Panin: Bible Numerics
*Bonnie Gaunt: Jordan-the Promise Fulfilled
*Bonnie Gaunt: Stonehenge and the Great Pyramid
#Yachov Rambsel: The Genesis Factor
+E.W. Bullinger: Number in Scripture
+E.W. Bullinger: The Witness of The Stars
Richard Amiel Mcgough: The Bible Wheel
Ray Comfort: Secret Facts in The Bible
^Jonathan Lawton Hammond: THE GOD-CODE AND CREATION

They've all written books with pertinent information regarding these particular topics of study, which are up to date, startling, and can be immediately
applied to scripture studies. Some of their theories you may agree with some you may not, but the numbers applied to scripture don't lie and need no interpretation. I strongly urge the

acquisition of at least some of these books to get started. Namely: The Genesis Factor by Yacov Rambsel, Jordan-the Promise Fulfilled by Bonnie Gaunt and THE GOD-CODE AND CREATION by Jonathan Lawton Hammond.

There are also a number of websites you can Google to go along with this reading material but please be careful when surfing the web: men like Jeff A. Benner, Brad Scott, Mark Biltz, Jim Staley and Chuck Missler. Their websites include: AHRC (ancient Hebrew research center), Wild branch ministry, El Shaddai ministries, Passion for truth ministries, Koinonia House and 119 ministries- all of which can be googled for links. I do, once again, caution against the use of any material if not approached humbly and fearfully. This is a study of God's Holy, perfect, Word so do not abuse! Let's get started.

If you can imagine an overhead projector where transparent papers, with labeled diagrams on them, are laid consecutively one on top of the other, you can begin to perceive the layered phenomena the bible contains. It would be much like a health class project where the teacher would apply one transparent paper with bone structure pictured on it, laid on top of another with coronary arteries on it, then another with soft tissue on it, and fill in the blanks with each corresponding paper laid down, culminating with a detailed picture of the human body with all of its functioning systems outlined.

This is the majesty with which God's Word is given to us by way of the Hebrew and Greek languages. It contains multiple layers that go beyond the surface text and journey deeper into the very Wisdom of God! Hear me, though, when I say the surface, written text is the most vital aspect of the Holy writ, thus he truly is the Alpha and Omega, the Aleph to Tav.

Because everything found beneath the surface, through Holy Spirit revelation, points directly back to the written text. You won't find a different story told in these deeper dimensions of God's Word. You will find them to house the same story of redemption through Christ the redeemer, only from different angles and a broader understanding then you're probably accustomed to. All which supply confirmation and point to divine authorship!

There is a plethora of spiritual insight to be gained; beginning with just the surface text, that emerges once you start cross-referencing verses. For example: we've all read the "symbolism" employed when we overlay the call of Abraham to sacrifice his son Isaac on Mt. Moriah with Yahweh's intent to sacrifice his very own son Yahshua on the stake. We've read it but did Wisdom discern it for you? Gain insight!

We see how Abraham's journey is a foreshadowing of Messiah's sacrificial offering centuries later on the very same mountain! *Lay a sheet down.* We also see the blood at Passover symbolically foreshadowing the blood at the cross. *Lay a sheet down.* We see in the Torah, sacrificial laws underlying the necessity of an unblemished offering, hence, the person of Christ nailed to the tree! *Lay a sheet down.* Can you see this through the overhead projector now?

What about the redemption from bondage through Moses foreshadowing the Redeemer to come? *Lay a sheet down.* And how about Noah and his God given means of escaping the deluge to come, and the atonement Christ makes available through his shed blood? *Lay a sheet down.*

They all paint a collective picture of the love God has for us through the arrival (prophecy fulfilled *lay a sheet down*), impalement (prophecy fulfilled *lay a sheet down*), resurrection (prophecy fulfilled *lay a sheet down*), and one-day return of Gods' Holy Anointed Son!! There are many more examples of these, sprinkled throughout the written text that a good thorough study of his Word reveals. It is miraculous! And below the surface there's more.

Let's begin in GENESIS at Aleph (the Beginning); bearing in mind that the Hebrew alphabet and language possesses multiple meanings and definitions. First, let's look at the meaning of names and take the genealogy from Adam to Noah. If you've seen this before good, let it strengthen your faith! If not, drop the scales and learn! It goes as such (and thank you Chuck Missler):

Name	Hebrew translation
Adam/Adomah	man
Seth	appointed
Enos/Enosh	mortal
Kenan/Cainan	sorrow;
Mahalllalel/Maleleel	the blessed God
Jared	shall come down
Enoch	teaching discipline;
Methuselah/Mathusala	his death shall bring
Lamech/Lemech	the despairing
Noah/Noe	rest or comfort

Read it, it's absolutely astonishing!! The plan of salvation, (of man by God himself), is spelled out through these men's Hebraic names in the very first chapters of the bible. Scripture is saturated with these types of examples once you factor in gematria, codex's, scriptural numerology, zodiacal witnesses and the like! The Word explodes with brilliant, majestic glory (I don't know if zodiacal is a word).

It's perplexing in light of the revealed, to see man search to and fro for the "meaning of life". It's amazing how man will point his Hubble telescope to the farthest reaches of the universe in the desperate search for intelligent life when it, "the meaning of life", has been in his possession for, at the very least, the last five centuries.

NASA, SETI and your pioneering frontiersman are all over-looking the treasured possession for a "new discovery"! But as Solomon says in ECCLESIASTES "there is nothing new under the sun" (ECCLESIASTES 1: 9), including co-incidences, as everything under the sun is the result of divine design!

So as the atheists purge onward, searching the heavens above for an inkling of life, which their evolutionist thinking dictates must happen, I'll be thankful for the treasure we already possess! Once we begin to open up the knowledge scripture houses, scientifically, we start to see that science really is of God. Man has his *stolen* science (fruit of knowledge) while scripture contains God's pure science (God's Wisdom). You decide which fruit to bite! A healthy read of PSALMS 8 renders to God what is God's.

Now before moving on, let's take a look at another example of God's mesmerizing Wisdom! It has to do with universal ratios (a stalwart in the scientific community) and the numbering of the Hebrew and Greek languages known as gematria. What we find is the signature of divine authority stamped upon his entire creation in undeniable fashion!!!

Here is a teaser nugget from the two books I suggest you acquire *first* to begin your study, "The Genesis Factor" by #Yacov Rambsel and "Jordan, the Promise Fulfilled" by *Bonnie Gaunt. And remember this, there are no coincidences, and "there is nothing new under the sun"! **NOTE**: I do not ascribe to the collective salvation conclusions that Bonnie Gaunt does; and in light of the book of Enoch, I do not hold mt. Hermon to be holy. I am merely extracting the mathematical intricacies she and others have unearthed in this geographical location.

The distance from earth to the sun is ***93**,000,000 miles. Mount Hermon, on Israel's northern border, peaks a ***93** mile long Mt.

Range *9,300 feet above sea level. No biggie? Keep going! The (gematria) values for "God is spirit" (JOHN 4: 24) ***93**0, for "brightness" "*of the sun*" (ACTS 26: 13) ***93**0, for "love" ***93** and for "to save life" (GENESIS 45: 5), you guessed it, ***93**!

Again, this is no big deal right, just another co-**incidental** happening right? Let's keep going and see just how non co-incidental this all is! Lets move on to a completely different book and method (codes), and see what is revealed. And remember, this is but a hint of the knowledge scripture houses within the Hebrew and Greek languages.

Did Adam and Eve die the day they ate the forbidden fruit? Spiritually yes, but physically "#Adam lived #**93**0 years", and that is #**93**% of a thousand year day and "a day is like a thousand years" (2 PETER 3: 8). So they did in fact spiritually die the day they ate the fruit, just not according to time as we neither perceive it nor understand it to have been in antiquity!

And when you begin with GENESIS 1: 24 and "#starting with the third letter in the first word, counting forward every #**93**0 letters spells Adam"! And within these coded scriptures we find that “#adjacent letters at the same e.l.s. (equal letter spacing) spells Eve, Cain, Abel and Seth”!! Evolution, ha!!

Add to that the degree necessary to view a rainbow, 42 degrees, and a double rainbow, 51 degrees and what do you know? **93** degrees (52+41) is, “coincidentally” the number God chose as mans’ necessary positioning to view this “sign” given to post antediluvian man by God! One degree closer or farther away and the rainbow disappears, can you say, “Let there be light” and the color prism, man?

There is an absolute spiritual connection or link between the number 93 and spiritual goings-on. One other interesting note: scripture declares Jerusalem to be the inheritance of Jacob, hence Israel, exactly **930** times (credit Jack Van Impe).

The findings when comparing the gematria and code values of these two distinguishable languages, the Hebrew and the Greek, with the mathematical-equation language and galactic dimension formulas and the ^angstrom measurements of the light spectrum used by science, are absolutely astounding, enlightening and life altering!! Please acquire all of the books mentioned above, beginning with these three: "The Genesis Factor", "Jordan-the Promise Fulfilled" and "THE GOD-CODE AND CREATION".

But moving on, here's a condensed lesson on the subject, which is laid out in much more detail in the reference materials named earlier and the above named books. Or, again, you can search the web for the materials to undertake this endeavor.

The act of assigning a numerical value to these alphabets is not a modern invention, it is as old as the written text themselves and it goes something like this; each **letter** in these alphabets hold a corresponding **number** value. Remember back in chapter one where I told you numbers and letters, the ability to calculate and speak, are of the spirit realms? This is its' applicability. I have italicized the Greek merely for differential purposes, thus:

Hebrew/*Greek*

Aleph/*Alpha* =	01
Bet (Vet)/*Beta* =	02
Gimel/*Gamma* =	03
Dalet/*Delta* =	04

He/*Epsilon* =	05
Waw/*no Greek equivalent* =	06
Zayin/*Zeta* =	07
Het (Cheyt)/*Eta* =	08
Tet/*Theta* =	09
Yod/*Iota* =	10
Kaph (Khaph)/*Kappa* =	20
Lamed/*Lambda* =	30
Mem/*Mu* =	40
Nun/*Nu* =	50
Samekh/*Xi* =	60
Ayin/*Omicron* =	70
Peh (Feh)/*Pi* =	80
Tsadi/*no Greek equivalent* =	90
Koph/*Rho* =	100
Resh/*Sigma* =	200
Shin (Sin)/*Tau* =	300
Tav (Thav)/*Upsilon* =	400
No Hebrew equivalent/*Phi* =	500
No Hebrew equivalent/*Chi* =	600
No Hebrew equivalent/*Psi* =	700
No Hebrew equivalent/*Omega* =	800

I realize this is a rudimentary outline but it gets you acquainted. Again, the aforementioned accompanying material lays this all out in much better detail; so get them if at all possible! And the Internet is "teeming with fish", *figuratively* of course.

The remarkable structure of God's entire creation will begin to emerge as you apply the number values to the text. The universal ratios employed by nature and the scientific community align themselves as the perfect numerical match with the gematria and coded values of the text found in scripture. You'll see!

And know this; the numeric values are just the beginning. The pictographic ancient Hebrew texts delve even further in to the meanings of these letters, hence the words, hence the message, hence The WORD of God! Each letter, in and of itself, tells the story of creation and redemption; quite comical as there's no mention of natural selection, survival of the fittest or big bangs and evolution. Okay,

You can do an exhaustive study by learning the Hebrew and Greek languages and applying gematria and codes yourself, but learning two languages is a daunting task. *Oh the monotony*! Then you could go back to school and become a certified scientist.

After that you could shoot for diploma in physics. You might as well go for microbiologist too, all the while keeping your nose in the Word! Remember, that is the key to this!

Or you could do what many are doing; you can buy these books by these authors, get your eyes on the treasure in scripture and learn. Humbly and fearfully, learn! As you proceed, bear these numbers in mind as scripture tags them as foundational to God's creation: ^/*3, 7, 10 and 12. 40 is a number ripe with meaning as well!

David reigned over Israel forty years (1 CHRONICLES 29: 27) and Solomon reigned over Israel forty years (2 CHRONICLES 9: 30). Joash reigned in Jerusalem 40 years (2 CHRONICLES 24: 1). Moses was forty years old when he defended a fellow Hebrew and killed an Egyptian slave-master (ACTS 7: 23-24). Moses fled to and remained in Midian forty years until Christ the burning bush beckoned (ACTS 7: 30). The nation of Israel wandered the wilderness forty years (NUMBERS 32: 13).

It rained on the earth forty days and nights (GENESIS 7: 4). Goliath taunted Israel forty days (1 SAMUEL 17: 16). Yahshua Jesus fasted forty days and nights (MATTHEW 4: 2). Jonah warned Nineveh of impending doom to arrive in forty days (JONAH 3: 5).

Moses was on the mount of God forty days and forty nights (EXODUS 24: 18). The Israelites returned from exploring the land of the Canaanites at the end of forty days (NUMBERS 13: 25). Joshua was forty years old when this exploration took place (JOSHUA 14: 7). Elijah traveled forty days and nights to be met by God at Mt. Horeb (1 KINGS 19: 8). Egypt was made desolate forty years (EZEKIEL 29: 11-12). Messiah's post resurrection ministry, which focused on the kingdom of God, lasted forty days (ACTS 1: 3).

Forty men, over a long unfolding of time, wrote the Bible; and on a personal level, God waited until my fortieth year of life to move me to this writing. I do not know the importance of the number forty but I know scripture places a definite significance upon it.

Honestly, all numbers are essential to creation: from the tiniest of particles to the largest of stars, from the smallest integer to infinite calculations; and scientifically verifiable as the building blocks of light, water and oxygen: “THE GOD-CODE AND CREATION” by ^Jonathan Lawton Hammond is superb and very detailed in scientifically matching “scientific” numbers with God’s creation.

^DNA, elements, molecular structure, particles, atomic numbers, proteins, aminos, immune responses and man’s very body temperature are all pertinent to our life on this rock, in these jars of clay. Again, this is a recent discovery for man, yet known from everlasting to everlasting by our majestic Creator; Jesus is his name!

The absolutely miraculous gift of life, and the God given methods by which life is sustained, is of our awesome Creators’ magnificent Wisdom!! The egotistical smugness of the atheist evolution theory thumpers will be put to shame once you shine the “**light** of the world” (JOHN 8: 12) on their ideology!

They stole science and perched it opposite God’s Word. Take it back, put it in its’ proper place and teach your children the fruits that such an endeavor will bear! That’s right; counter the drek pumped into them at school with the Truth of their Creator Yahweh God, El Shaddi! Simply seek teaching over training, the **Word** over the **world**. "Shema Israel"!

One area that will open your sleepy eyes even more is the theme of astral knowledge. It's where the astrologers and astronomers coagulate and coalesce with their like mindedness! However, they glorify the stars instead of the one who placed them there. Do not do this!

It's the ancient form of future telling. But what man misses the mark with is the denial of the much bigger declaration the constellations are witness to! Along with signifying YaH's appointed feasts (our time clock for prophecy – which we are commanded to know) they speak of the same story of man and his need of redemption and the Messiah who accomplished this task. "The Witness of The Stars" by +E.W. Bullinger sheds a more thorough light on the subject, just be sure to have scripture opened to (JOB 38: 31-33 and 9: 6-10) and (AMOS 5: 8) and (PSALMS 148: 1-6) and (JEREMIAH 31: 35) among many others when you diligently pursue this undertaking.

Do you know that when we look up to the heavens we are reading scripture? When we star gaze, we are reading a different version of the same creation scripture depicts. Individually, horoscopes focus on the smaller picture of our lives, but collectively they paint the same picture that scripture teaches. Exit the state of being trained by the world and enter the teachings of our maker- Aleph to Tav, Jesus of Nazareth.

The essence of everything, in the final analysis, is this: The Holy Bible, Holy Scripture, the Word, is a directive and instructional manual for all creation. All of creation, from the farthest star to the tiniest molecule; from the biggest celestial body to the smallest blood cell, from the clouds and wind and rain and volcano's and earthquakes to the vegetation, from the birds of the air to the creatures of the sea to the creatures on land, are all obedient to the Word; obedient to TORAH, obedient to the bible!

All of creation is obedient to this book man possesses. Everything is obedient to, and held in check by, God's Holy Word! As far as I can tell, there are only two creations disobedient to his Word, the fallen angels and fallen man. Man, wake up and get on the right path. His loving Grace makes this possible, but if pride is your choice, you make an uninformed choice. The bible is an instructional manual, the history of mankind, the story of mankind, and the story of the redemption of mankind, the means by which that salvation is met and, on top of that, it binds all of creation together. Be trained by man or taught by God, man. Read Psalm 119, with its' acrostic structure and begin to comprehend David's thoughts and heart when reading it.

Just know this; the writing we possess is much more than the words on paper, much more than a story about Jews and gentiles and Israelites and so forth. It's the story of all creation: in multiple dimensions and with multiple directives and instructions. It is the Way, the Truth and the Life of everything that has ever been, is and will ever be so, WAKE UP! Every time Christ declares, "I tell you the Truth", do you know what he is teaching us? The TRUTH; the multi-leveled, multi-faceted TRUTH! Grace be with you my brethren.

Chapter 6: Theological Truth

The essence of everything perceived can only reach us by way of our senses. They funnel through our eyes and ears before being filtered through the vast network of censors, receptors and nerve endings housed in our craniums and spines, through which flow all neurological activity. This hard drive is no easy task to duplicate either; in fact it is miraculous and utterly impossible to duplicate! And the hope of ever claiming to create it ourselves is utter folly!

Our mind *also* houses the gate with which the spirit realms inter act with us personally! Through its' highly complex physical micro network the human mind observes, retains, deciphers and evaluates massive levels of vital information while simultaneously sustaining every other complex bodily functioning system! Without a plug!

This is an absolutely miraculous feat to accomplish yet this is "only the tip of the iceberg"! In the human mind (the seat of emotion and reason) is where the physical and the spiritual collide! This is where the fight for your opinion, spoken through your tongue, thus your life, occurs!

Man, regardless of how intricate and advanced his new "state of the art" hyper-speed supercomputers may become, cannot duplicate this spiritually interactive feature! Physical matter we can manipulate and replicate; where do you think the new state of the art, high-tech cameras get their capabilities for clarity? From the study of the God-created human and animal kingdom eyes! All technical aspects of image display emanate from the duplication of the blueprint, of the blood and breath creatures' eyes, that God created.

The study and dissection of the God-created human ear is where our audio enhanced experience of life emanates from as well! These are only two examples of man stealing and calling it his own. But take away the eyes and ears that God has created and man has nothing to pattern his "inventions" after, nor the ability to do such patterning!

What about the mechanical design of pumps and hydraulics? The God-created heart and diaphragm are the pattern man copies, and then calls his discovery. I'd also like to point out a little feature about the human heart; it doesn't really pump blood throughout our bodies, it's more of a regulator to a natural occurring current powered flow.

What about image, sound, information and calculation devices called computers? You got it, stolen from the dissection, testing of and manipulation of the God-created human mind and body. Man's invented the means but who provided the raw material and know-how for such inventions? God! But after it's booted up, who controls the content?

Our computers, like their predecessors TV and radio, have become a conduit of influence from the spiritual dimensions, but they cannot discern or interpret Holy Spirit Wisdom; they, drum roll please…….compute! It's a one directional relationship and that direction is pointed at us, so be wary. In the end they are still just a plugged in or battery operated, mechanized "contraption" with plastic, silicone and metal parts.

No matter how hard we strive we can only build machines that collect data, run scenarios, spit out feedback and obey its' programming! They have the ability to equate and analyze as programmed but not rationalize or reason about the presence of spiritual elements, as man can! As I stated, we can mimic many attributes of the human mind but absolute duplication is impossible!

An interesting side note to our creative abilities and our innate nature for creation manipulation is our ability to discard, at any time of our choosing, any machine that disobeys our programming of it; until the "singularity" is reached anyway. I am so thankful God doesn't treat his creation, man, as man treats mans' creations. It may appear as though he discarded or "threw" us away by casting us from the garden, but remember how he then came to save that which was lost. For now let's return to the marvel of the human mind.

Sure, man has valiantly attempted to create its equal but an in-animate imitation is as far as our knowledge can take us! We've tried, but to no avail! Our synthetic, computerized world can only go as far as the next computer program allows us to go! Apparently we have to wait for technology to "catch up" to mans' notion of what is science fictionally possible.

But this is not so in God's creation! As long as God wills it, creation moans in anticipation for its end, this due to man and his fall! And as long as we remain in our fallen nature what do we do? We imitate God and manipulate his creation, all the while denying his existence! Putrid! This is precisely why his wrath is soon to be poured forth again!

And this is also precisely why the urgent necessity to abolish evolution at its' dark core!! We're running out of time fellow man! I do not declare this with a Chicken Little "the sky is falling" attitude, that would indicate an unfounded, unsubstantiated claim! No, it's actually much more dire than that!!

Scripture declares we are on the pinnacle of, if not already in the midst of, a spirit induced, life altering change on this rock we call planet earth! There are many prophesies that have already come to pass, many currently being fulfilled and more yet on their way! This should not be taken lightly! We have reached an age of closure and the times bear witness to this!

Thus the necessity to get to the deeper levels of spiritual Truth contained in scripture through a focused, purposeful study! It is vitally imperative as the scales begin to fall to be Holy Spirit guided through the minefield of modern life!

If you've done the study recommended in the previous chapter, then you can now begin to see the multiple layers and dimensions to God's Holy Word, yes the sword of Truth! If you are now prepared, good! You will need this Wisdom to withstand the days ahead! This is no fairytale or myth or legend, it is the reality of this modern life!

The desperate need to affix our heads on straight and leave the rebellious mindset of psychology and evolution is upon us! The only viable strategy is as scripture declares; the donning of the Word of God; yes the double-edged, blood and marrow-separating sword of Truth!

Once scale-less your heart, mind, soul and strength will detect that this is the overall best qualifying description of what scripture is, the sword of Truth! It's a spiritual weapon! When Messiah was tempted in the wilderness he didn't respond with reason or debate, he responded each time with "it is written", so we have to do likewise. And remember, satan tempted Christ with-scripture.

It is the weapon of choice for our Creator, his Messiah and his Holy Spirit! It is the only manner of attack, for man, sufficient to ward off the folly of the adversary, the principalities of the air and even man's trainings! And that is the order in which our foes line up to attack, which brings us to the crux of this matter!

As my attack on mans' evolution training is rightly justified, it too is only a partial slice of the much bigger picture! There is a spiritual battle raging in the undetectable spirit realms! But don't be fooled by an inability to see it, it's as real as the breath we breathe and the sound waves we hear, which we can't see either!

My point is this; don't foolishly deny that which is not perceived through our senses because the battle rages nonetheless! The war is over the very souls of mankind so do *not* presume to be excused from the fight! Its age old; it's been raging since antiquity and the ancients fought the same battles!

The dangerous difference is that the thought that they evolved from monkeys never even occurred to them!! They, the ancients, were entrenched and aware of the spiritual seducing influences upon man! Do not allow the deceptions of modern man to strip you of the knowledge and awareness of, or the weaponry needed to fight the battles of, this war!

Deception through stealth is the adversary's most piercing arrow! In any warfare, if you can *persuade* your enemy that you're not even there you've already won! Well, man, to satan and his army we, oh man, are that enemy! We are loathed from conception, hated at birth and pursued until death!

Satan, Lucifer, the devil, the adversary, the father of *lies* wants you to believe *he* does not exist (*liberal theology*), that our Creator YHWH does not exist (*evolution*) and everything that does exist is just a big celestial cosmic coincidence (*psychology*)! Do not be *persuaded*!

As we all gasp in horror at the images and sounds TV depicts, through current **n**orth-**e**ast-**w**est-**s**outh stories, filter them through the knowledge of this truth, satan wants you dead! When psychological mind manipulation through evolution prevails, spiritual death is imminent! His desire is that we all perish! And he will stop at nothing to accomplish this through any and every means allowed! Arm yourself!

And understand this man; God has given the celestial beings, angels and demons, much more room to operate within our physical realm than we may be aware of! God has allowed angelic beings more intrusion in to our lives, and more power to influence our lives, than *we* may think is fair. Our thoughts (psychology) on such matters, however, don't factor in to the equation. Does that strike at your heart, man? So what?

Do you caress the notion that God owes you an explanation? Or do you think that "*he would certainly never do that; give the angels and demons the right to influence us with out our consent*"! Yes he could and did do that and still remains right and just, you know why? He is El Shaddi God Most High!

This is all his creation, his concept, his doing and his domain sustained by his might, his power and his glory! Held in check with his standards, his regulations and his Torah Word so stop the murmuring in the wilderness, and sing praises to his name!

Praise God and comprehend this; the angels and demons and spirits are created as well, they can only go as far as the Creator allows them and not an inch further! Read JOB 1: 6-12. So if you "feel" violated know this, God has allowed it! Not to dismiss us but to draw us back to him! Read JAMES 4: 1-16. Just bear this in mind, this "reality" or dimension we inhabit is only one tree's experience-the tree of the knowledge of good and evil. Everything perceived in this dimension of reality is based on polar opposites, yin yang, light and dark, thus good and evil. Our whole existence is housed in but one tree.

With this firmly established read EPHESIANS 6: 10-17 and do as it states! Pick up the "sword of truth", the "shield of faith", the "helmet of salvation" and the "breastplate of righteousness" and adorn yourself with them! As I've stated repeatedly, scripture declares time is running short so don't delay another moment!

You're not from an ape, not a descendant of monkeys, not found in a chimp's lineage and not an orangutans' offspring; those are doctrines of demons and seducing spirits! **You** are created in the image of the mighty living YHWH El Shaddi! So suit up with the armor of God, purge the evolution mindset from among you, and seek the kingdom or when you die, you will die in your sin! Because even a miniscule agreement with evolution is a subtle form of denouncing Christ!

As long as you hold evolutions' idea as remotely possible you forfeit any salvation or redemption you might have otherwise attained through the shed blood! Yes the sacrificial blood of the slain Lamb of God! Please don't fall victim to "My people are fools" (JEREMIAH 4: 22). Arm yourself! Or you may well one day hear "depart from me, I never knew you"; all the while believing yourself to be a good Christian.

Arm yourself with spiritual battle fatigue for it is a spiritual war that you are in! Fight from your knees with fear and trembling wielding the very sword of Yahweh God! Its' multiple dimensions should verify its' power, supremacy, sovereignty and its' all surpassing glory! Just to clarify; your fear should be of "him who, after the killing of the body, has the power to throw you into hell. Yes, I tell you, fear him" (LUKE 12: 4-5), not the battle itself, nor its' participants!

So how do we fight this war? How significant is our part? How do we throw down with what we can't see or hear? How do we know when to take cover? How do we know which direction to point the sword? What assurance do we have of victory? Christ, who makes EPHESIANS 6: 18 possible in the first place!

Christ Jesus/Yahshua Elohim is the only answer for all these questions! The Messiah is the only way for he is "the way, the truth and the life" (JOHN 14: 6). And, now I'm gonna really open up a can of worms, he's not the Jesus of 12-25 fame! That's just gross history manipulation! The Jewish Messiah born in Bethlehem did not arrive in the cold Gregorian month of December, although he was probably conceived around this time.

If you are of the mind set that Christmas and Easter would have had any meaning to that man tortured on that stake two millennia ago, change! Change now! This is a hedonistic, pagan worship, nature worship, sun worship, and equinox and solstice worship and fertility worship mind set! It is wicked creation worship instead of Holy Spirit filled Creator worship. It is idolatry.

Here's the irony of it all, which contributes to the confusion of thoroughly "modernized", thoroughly "evolutionized", thoroughly "manipulated" and "liberally theologian-ized" man; the body of Christ (the church-bride) not only accepts these Christmas and Easter ideas, it cultivates them! It's just mind boggling to me how quickly the seeds of weeds take hold and sprout! Then again, we were forewarned of the tares, were we not?

We have a jolly ole chap called Santa Claus representing the birth of the Messiah on December 25th every year, which wasn't even his birthday, in conjunction with the same time the marketing and advertising executives start churning out the "holiday season" banter. Folly!

Then come spring, in the stead of YaH's appointed feasts Passover (blood redemption) and the Resurrection (first fruits), we have this bunny rabbit, a picture of pagan fertility, representing the tortured and resurrected Messiah? Again my mind goes huh? That's-but-should-what-huh? This is ironic, wrong, a twisting, nowhere to be found in scripture and an all-indicting fruit that bears witness! Arise from your spiritual slumber church; time is dwindling!

The adversary sowed the seeds in Freud, Pavlov's and Darwin's heads and in turn they willingly sowed the seeds in mankind's head and poof! Look at us; we can't even discern our error in this false idol worship! These are but two examples of the spiritual battles facing us and how quickly false religions and traditions can grow.

We acquiesce and surmise in to existence a Santa Claus and the Easter bunny? Have you ever taken the time to decipher the error of three days equaling 72 hours; contrasted with the training that Messiah was crucified by 3:00 pm on Friday, and then resurrected on Sunday morn? Do the math. Scripture declares one timeframe while religion teaches another entirely different timeframe. Deceptive and mighty powerful is the adversary, so wake up man as the deceptive manipulation comes at us from every direction in this spiritual battle! Again, do not be persuaded lest you have to endure hearing "depart from me, I never knew you". (Remember; in MATTHEW 7: 21-23, Christ says this to people claiming to be Christians).

Now literally, so how do we fight this war? Christ- ! How significant is our part?

Christ- ! How do we throw down with what we can't see or hear? Christ- ! How do we know when to take cover? Christ- ! How do we know which direction to point the sword? Christ- ! What assurance do we have of victory? **Christ -** !

See; don't ever take his name in vain like the Hollywood crowd habitually does, take it seriously, very seriously; cherish it! It is the name by which "every knee" both in heaven and on earth "will bow"! I do not care what the fallen angel mouthpiece "entertainers" think, I care not what the agenda pushers think and I don't give much thought to what the world thinks! I strive to only care what Messiah thinks!

God's Word is the barometer by which all creation will be judged, not the entertaining, agenda pushing worldly folk! So stop trying to "keep up with the joneses" every "holiday season" as this only renders you with a spirit of stupor! Much more vigor is required for this battle we're in! In chapter five I commented on the significance of PSALMS chapter 119 as a whole. In this spiritual warfare we are in, now would be the time to read, digest and fall in love with this particular chapter of PSALMS. Read and comprehend and own what, exactly, is on and in David's heart as Christ is Aleph to Tav, the fulfillment of that longing.

There are many areas of our lives affected by these multi-pronged attacks by the adversary and we cannot win these battles in the flesh! We are incapable of defeating any of the fallen angels' army while housed in our carnal vessels! If we try, our lives then become a futile, halfhearted attempt on our behalf, to somehow show ourselves worthy of the treasures in heaven, while living worldly on earth!

God sees right through our half-heartedness and the adversary sees right through it too! They are spirit whereas we, man, are spirit and flesh! The book you hold in your possession, scripture, is spiritual in nature, spiritual in application and spiritual in origin and the only means given as necessary for this war, through Holy Spirit guidance!

Adapt! How? Christ! Adapt your discipline and fighting style and technique to mirror that of our makers'! Slay the fakers with the Makers' Word! Eliminate deception with Truth! What fakers you may ask? The ones who've got it all figured out with their fake, phony and false trainings, both inside and outside the church.

This includes but is not limited to evolution, atheism, psychology, the new age movement, angel worship, alien worship, scientology and compromised, worldly, liberal theology Christianity! Any and I mean any consumption of this yeast must be avoided at all costs! This modern, evolution soaked theological sponge is not to be tasted by those who make up the true body of Christ! With proper Holy Spirit discernment avoid all forms of false and misleading trainings! They are doctrines of demons.

As Christ the Messiah declared, "He who is not with me is against me" (MATTHEW 12: 30)! Hear the voice calling in the wilderness, man! Hear and obey if you hope to ever see the kingdom of God! None of what we are discussing here should ever be taken lightly!

For those who *choose* to handle these matters lightly, they will be shown the lethal error of their liberal theology based assumption! What assumption? The one that assumes a loving God would never allow a portion of his creation to be cast asunder! You know; the "man is basically good" crowd.

They regurgitate things like "I never killed anyone". They live in a shell with "I'm a good person" as their defense. They make claims like "many roads lead to heaven" and "we don't view the bible as an historical book" or "we don't take scripture literally" or "the bible was written by man". Be wary of such manipulative babble!

Although they may very well be sitting in the pew next to you, they speak the language of the gurus! They speak in unison with the evolution ilk and they act in tandem with the entertainer's guile! They think they are "square with the big man upstairs". Thus, unknown by them, their fruit identifies what type of tree they are!

The fact is they belittle the urgent need of repentance, believe works and charitable contributions are their ticket to heaven and worse; they interpret the Word of God with their "feelings" and "emotions" rather than divide it correctly! All of these notions are; false and misleading, dangerous, from the spiritual realm, believed by many and lead to death!

Do not ever forget this is a spiritual war we are in here, and plopping the behind in a church once a week is good but not a deed worthy of redemption. It never was! Please don't misconstrue my implication here!

Yes, as the apostles teach: it *is* good and necessary and commanded to assemble together to edify and uplift the preaching of the Gospel and to sustain the body of believers, but attending a church service has no redeeming value in and of itself! The Word saves; not perfect attendance! If you are parked in front of a liberal theologian spewing out false doctrine, perfect attendance can't undo the damage being inflicted. Do the proverbial fly and the window shed some light?

I've known people who sit in the pew every Sunday as deacons and elders, then turn around Monday through Saturday taking God's sovereign name in vain, all the while listening to guttural music, which also slanders God's Holy name, pumping out of their radios as they sing along. This lack of continuity also bears the fruit of the tree as it carries the stench of liberal theology marinated with evolutions' innuendos.

And this should also clue us in to the multiple arenas of battle the enemy spreads his spiritual tentacles throughout! There are those who deny the Christ outright (evolution), the ones whom this book you are reading finds its' premise in, and there are those who are a little more discreet with their distaste of the Gospel (liberal theology)! However,

There is much more to this spirit war we are engaged in. It is one thing to identify the fruit outside the body but to properly and accurately identify the fruit of those within the body takes Holy Spirit discernment! That is why attempting to undertake this battle in the flesh, through mere interpretation, will only lead to division, derision and a corrosion of the faith!

Fight it in the spirit not the flesh! Our carnal faculties severely limit our ability to discern, period! So don't mess around here; one error could heap burning coals back on to your own head! Do precisely as scripture commands, Word for Word, "line upon line, precept upon precept", yes-literally! Do not interpret the Word; divide it correctly! Foreign words and languages require interpretation whereas scripture requires obedience.

"Love Yahweh El Shaddi God with all your heart, all your soul, all your mind and all your strength" and "love your neighbor as yourself", regardless of how you "feel" about them, for this is the "sum of the law"! And this is the "sword of Truth" so wield the weapon wisely, it is a gift remember? Let us now do some wielding! And know this; reversing the order of these two commands, the norm in today's day and age, leads to a bevy of trouble.

It forces one to cling to the "golden rule" yet negates the beauty and urgency of the command which precedes it; you know the love God first command.

This mindset makes it all too easy to disregard the front half of the book; the Torah and the prophets, those cumbersome commands and feasts we've convinced ourselves were for the Jews only. But when you actually read what the text says you will realize that God, himself, calls them the Feasts of the LORD, not the feasts of the Jews.

Whoooosh, did you hear that? That was the sword of Truth whizzing over your head so you better start paying closer attention in this cosmic war!

Oooh oooh oooh?

Chapter 7: Weighing Words

"In the beginning was the **Word**, and the **Word** was with God, and the **Word** was God. *He* was with God in the beginning. *Through him* all things were made; *without him* nothing was made that has been made. *In him* was life, and that life was the **light of men**. The light shines in the darkness, but the darkness has not understood it" (JOHN 1: 1-5).

Hearken back, man, to the unsealed Hebrew scroll of his creation in the opening words of GENESIS 1: 1 that reads "In the beginning God created the heavens and the earth"! And read right on through GENESIS 1: 4 where God says "Let there be light"! Alpha-Aleph!

Now read JOHN 1: 1-5 again, only this time remove the dross, see God's scientific Wisdom at work and remember all of the information and knowledge acquired from the recommended reading material outlined in chapter 5. If you haven't done that yet that's okay, just reread this chapter again once you have.

If you utilized these tools my hope is that you view these verses of scripture in an expanded light now, as illuminated by God's Holy Word, instead of someone else's interpretation of it! A physically and spiritually literal reading of El Shaddi's life giving **Word** truly transforms the heart of man!

It is the light spoken of in GENESIS, for as it says, "The **Word** became flesh and made his dwelling among us" (JOHN 1: 14)! And that **Word** that became flesh spoke these words: "I am the **light** of the world. Whoever follows me will never walk in darkness, but will have the light of life" (JOHN 8: 12). Omega-Tav!

Read the rest of GENESIS 1: 4, it says, "God saw that the light was good, and he separated the light from the darkness". The *spiritual* aspect of this began in GENESIS, literally! The *physical* aspect in the person of Yahshua/Jesus, literally! Alpha and Omega-Aleph to Tav!

The perfect **Word** of God set the heavens and earth in place by merely speaking them in to existence! Now that is truly "Unlimited Power", Mr. Anthony Robbins! And many sands of time later that very same **Word** allowed itself to be housed in the form of man to extend the olive branch of salvation to all of mankind.

God's all powerful; all knowing **Word** literally left his throne in the heavens above and willingly took on the flesh of fallen man! A finite flesh bound by time and marked for death! Can our finite minds even grasp the full implications of this one act of selfless love? JOHN 3: 16 is a blessing beyond the ability of our carnal minds to fathom!

God, in his infinite Wisdom; created man, commanded man, banished man, sustained man, became man and suffered mercilessly then died at the hand of man, all to redeem… man! Do you get it yet, oh man? I tremble in fear and awe of the one who made me when this really sinks in! Truly he is the Alpha and Omega! His Aleph to Tav love surpasses all understanding!

This is the whole declaration of scripture! This is the whole history of man written down for all mankind to see! All creation screams aloud of the glory of the Creator, so why won't you world? I'll tell you why. Your evolution, "there is no God", psychology laced thinking has altered your ability to perceive God, rendering you spiritually blind and dead. Remember, “Blinding science” (blience)?

But for those of you whose perception skills are still in tact, may God bless you! Even though I want you to realize the deeper levels of the surface text, I will once again remind you that the surface text is what all of the different layers of scripture point to! As scripture lays out the plan of redeeming man through the shed blood of Christ, the deeper layers of confirmation spell out the very same plan in uniformity.

As the Torah, the books of Wisdom and the prophets all point to the coming Messiah, the Messiah points us back to our Heavenly Father, God! Yahweh El and Yahshua Elohim are the Aleph and the Tav, the Beginning and the End of everything! Praise Yah!!

What does this all lead to? Words! The importance of words and the weight they carry is enormous with YHWH! “In the beginning was the **Word**, and the **Word** was with God, and the **Word** was **God**”. Do you see it yet? Does your spirit discern the importance of the words man speaks?

Words are plural, for man to toil with! **The Word** is singular and all encompassing! Thus where our words, and our ability to even speak our words, find their genesis! You must understand the vast volume of weight that the privilege of speaking carries! Do not take it for granted, do not handle it lightly and do not ever assume when partaking of this God-given, spiritually charged, privilege!

Read HEBREWS 4: 12-13 and PROVERBS 30: 5-6 to effectually gain vital understanding in this crucial area of separating our *words* from **The Word**! Read Deuteronomy 8: 3 and MATTHEW 4: 4 for life and en**light**enment while toiling with your own words!

Here is a quirky little example of mans' mincing of words: live (rhymes with give), until the electronic explosion we witnessed in the last century, was the antithesis of die, hence to live or die! Now live (rhymes with strive) also means the opposite of recorded, hence live or recorded. I do not yet know the ramifications of such subtle meddling with words; I only know it reveals mindset and a willingness bent toward word manipulation!

Here is another one: the designation prophet at one time defined a messenger of God, but now profit is the bottom line in relation to earnings! I realize they're spelled differently, but if you are not privy to context or seeing it written down and only heard it spoken of, which of the duel meanings would it take on? How about Adam and atom, or genes and jeans, or right and write or rite? Again, this subtle mixing and mingling of words is a capricious display of an evolution drenched, psychology lathered mindset!

You might be tempted to surmise that I'm merely splitting hairs or hares on this issue but please pause before coming to such conclusions! Return to **The Word** and comprehend the emphasis our Creator places on the speech that exits our mouths, which does entail every single word!

Remember, "out of the overflow of the heart the mouth speaks" (MATTHEW 12: 34). And again "The tongue has the power of life and death" (PROVERBS 18: 21). And once again "He who guards his mouth and his tongue keeps himself from calamity" (PROVERBS 21: 23). **The Word** spoke these words, so you should discern the weight and impact of every word we toil with to speak!

When the mouth goes around slandering and cursing it reveals a heart that is cursed and slanderous! When the mouth goes around damning everything in God's Holy name it reveals a heart that is damned! Witness the fruit and stop philosophically interpreting scripture, and listen to what the spirit of Truth speaks to your heart, literally!

Let's return to the words "sticks and stones can break my bones but words can never hurt me"! As previously eluded to, this statement is an outright lie we infiltrate our children's minds with at a tender, young age! It belittles the weight words possess and it's just not true! Let's look at what scripture says and compare and contrast these opposing views.

Right out of the shoot PROVERBS 25: 15 states, "A gentle tongue can break a bone"! So much for the decades long lie we've mis-informed our kids with! But to avoid any conjecture, allow scripture to speak:

“A lying tongue hates those it hurts”-**PROVERBS 26: 28.**

“Do not be quick with your mouth”-**ECCLESIASTES 5: 2.**

“the mouth of the fool gushes folly”-**PROVERBS 15: 2.**

“but a deceitful tongue crushes the spirit”-**PROVERBS 15: 4.**

“the mouth of a fool feeds on folly”-**PROVERBS 15: 14.**

And for reprieve,

“the tongue of the wise brings healing”-**PROVERBS 12: 18.**

“The tongue that brings healing is a tree of life”-**PROVERBS 15: 4.**

“Even a fool is thought wise if he keeps silent”-**PROVERBS 17: 28.**

“The mouth of the righteous is a fountain of life”-**PROVERBS 10: 11.**

As you can see this is but a mini list that doesn’t even scratch the surface, but I hope you get the gist of importance, which our Creator places on the mouth, the tongue and the words they form! To gain a full, more in-depth revelation in this one crucial life or death issue, read the entire book of PROVERBS, then read JAMES 3: 1-12.

We must understand the vital and crucial emphasis God places on the words we speak, so we can assess the words the evolutionists speak, to compare and contrast what is right and what is simply deception! We should be placing an extremely high importance tag on this topic!!

We believers had all better make it our top priority to fulfill our duty of stewardship as scripture directs, not as our "feelings and emotions" prompt us! And for the unbelieving among us I plead with you to, at the very least, consider a "momentary lapse of reason" when Truth interrupts your life! You never know when or **if** the opportunity will again present itself!

It is our sovereign duty to rigorously attempt living a life bound to scripture instead of conformity to the worlds "modern" views! Listen, evolution is deceptive and flat out wrong! Wrong thinking, wrong philosophy, wrong ideology, wrong psychology, wrong training and the wrong thing to be pouring out upon our children! I only hope it is not too late for retribution.

The diabolical wretchedness of the evolution training is that along with being wrong, it is most effective! Its' smarmy tentacles have dripped down through time and caused many to doubt the absolute Truth of God's Word, and doubt the existence of God himself! This alone almost begs God to pour forth his stored up wrath!

With our words bent towards the notion of "evolving" we have symbolically returned to the knowledge tree in the garden! "Their throats are open graves" (ROMANS 3: 13). Read all of ROMANS chapter 3, completely, for this to really hit home! The tree of knowledge is where the serpent planted the seed of doubt to begin with, as words from the throat of the grave enticed us unto death!

We fell for it then and we still fall for it today! "You will be like God" is as enchanting, alluring, tempting, desirous and lethal as it was in paradise. The atheist evolutionists might not even be aware of its' power to hypnotize, but they've fallen for it nonetheless! Or they actually believe it. Moreover, man is not Yahweh!!! Be "like God" or be obedient to God, hmmmm? Fallen man will always choose the former over the latter for one simple reason; "there is no one who understands, no one who seeks God" (ROMANS 3: 11). Verses 10-18 should be applied for context and clarity.

Here's a silly observation: if every man and woman were mute, could the fall in Eden have occurred? Could evolution have ever even been taught? No, neither of these two exchanges is possible without…words! So when God speaks in his perfect Holy **Word** about the words we speak, we better listen to what the spirit says! And be informed; oooh oooh oooh aaah aaah aaah wouldn't have cut it either!

What about numbers and how they apply in all of this? Good question, I'm glad you asked; Multiplicity. Every word we speak infringes upon multiple dimensions, the spiritual *then* the physical. As hard as it may be to conceptualize, words simultaneously burst forth on a journey in to multiple realms of reality or dimensions, one word at a time! They're counted, measured and numbered according to scripture!

As we speak them, our words issue forth both spiritually and physically. Multi-dimensional is this gift we house. To reiterate;

"**By your words** *you* will be acquitted and **by your words** *you* will be condemned" (MATTHEW 12: 37), is it sinking in yet? Weighed and balanced in right and just heavenly measure! Speech and calculation on his cosmic scales, according to his laws, precepts, statutes and creeds! Remember, "as sharp as a two edged sword"!

Bbbbuttt, as he always does, man tries to out smart God, out wit his Creator! How? Well, modern man has changed and "r***evolutionized***" his speech habits to not include biblical words and, in stealth like fashion changed their labels, thus their meaning. This is going to be harsh, but test the spirit! If you find this doinks a little bit read the aforementioned verses in PROVERBS and JAMES once or twice more.

Let's review some of the celestial adversary's deceptively subtle and cleverly cunning seedlings, via evolutions' psychology enticed verbal mindset! I call this beacon of word manipulation the label changing manipulist:

First off, the heavens became "space" or the "universe"

A drunkard became an “alcoholic”
Fornication became “shacking up”
An unborn baby became a “fetus” and
Fetus murder became “abortion”
Murder became “homicide”
Lying became “stretching the truth”
Homosexuality became “sexual orientation”
Child molesters became “pedophiles” and
Pedophiles became “victims”
Adultery became “a fling”
Infidelity became “irreconcilable differences”
Divorce became a “split up”
Character assassination became “satire”
Slander became “comic relief”
Gluttony became “fine dining”
Coveting became “keeping up with the Jones’s”
Stealing became “borrowing”
Cheating became “getting an edge”
Greed became “wealth building”
Pride became “self-esteem”
Arrogant became “confident”
Malice became “expression”
Gossip became “commentaries”
Pharmacia became “prescriptions”
Carnal lusts became “addictions”
Prostitution became “making a living”

Sexually depraved became "promiscuous"
Enabling became "tolerant"
Opinions became "truth" as
Truth became "subjective"
Faith became "a crutch"
Hope became "wishing"
Love became a "feeling"
Lust became "passion"
Stiff-necked became "persistence"
Disobedience became "short comings"
Remorse became "depression"
Abstinence became "out dated"
Contraception became "safe sex"
Double-mindedness became "bi-polar"
Out of control became "compulsive"
Families became "two income households"
Babysitting became "childcare"
Sparing the rod became "time-outs"
Spanking became "abusing"
Lewd behavior became "kids having fun"
Violent words and actions became "blowing off steam"
Evil became "good" as
Good became "evil"
Divination became "meditation"
Wants became "needs"

Privileges became "rights"
Desire became "necessity"
Emotions became "decision makers"
Children became "young adults"
Indoctrination became "education"
Education became "higher learning"
Angels and demons became "aliens" or "extra-terrestrial"
Humble became "insecure"
Meek became "weak"
Un-repentance became "independence"
Entertainment became "art" and
Life and art became "imitations" of each other,
Dependent became "needy"
Revenge became "vindication"
Entertainers became "*gods*"
Idolatry became "emulation"
Aged became "elderly" as
Elderly became "expendable"
Banks became "loan officers"
Usury became "credit"
Doctors became "health practitioners" as
Drug stores became "pharmacies"
Morals became "inhibiting"
Modest dress became "boring"
Women's rights became "women's lib"

Scantily clad became "sexy"
Sex became a "seller"
Taxpayers became "consumers"
Time became "money"
Prevention became "passé" as
Cures became "essential"
Emphysema became "COPD"
Diabetes became "1, 2 and 3"
Mothers became "working moms"
Fathers became "dead beat dads"
Unwanted pregnancies became "mistakes" as
Mistakes became "planned parenthood" and, tragically, any mention of sin became "omitted" from our speech completely, as sin became "human nature"!
And here's a little sneaky one; paganism, hence creation worship, became "evolution"

This list could go on and on and on! It displays "in your face" blatancy when you simply take the time to dissect it! They are guilty of trying to pull a fast one on our Maker, but all they're doing is pulling the wool over our eyes so the religion of evolution belief can continue to flourish! Please wake up man, and drop to your knees and repent before the coming wrath arrives; *universally* just, weighed in the balance, deserved wrath!

Take a moment and contemplate the ramifications of completely omitting the label of sin from our vocabularies. This is precisely how good can be called evil and evil can be called good. The barrier has been done away with, the standard has been relinquished and deception flourishes. I desperately want you to deeply ponder what I just said; sin is not called a sin anymore!

To call anything a sin is now insensitive, judgmental, narrow, intolerant, politically incorrect, uncivilized and just not nice. Listen, a sin is a sin people! Man will not get away with this twisting of words.

Here is another example of adversarial verbal success within our psychological being, which sheds light on our present mental state. It has to do with the evolutionist "programming" we have undergone at the beseeching of the entertainment industry! It is also indicative of the TV dictatorship I pointed out in an earlier chapter. Try as hard as you possibly can to not remember these! And when you realize that you cannot forget them, thank the companies responsible for these jingles/slogans:

“have it your way at B _ _ _ _ _ _ _ _ _g”!
“Mc _ _ _ _ _ _ _ _, I’m lovin’ it”!
“where’s the beef, _ _ _ _ _s”!
“C _ _ _ _ _ _ _ _ _ soup, mmm, mmm good”!
“O _ _ _ _-M _ _ _ _ has a way with bologna”!
“I am stuck on B _ _ _-a _ _ cause B _ _ _-a _ _ ’s stuck on me”!
“the best part of waking up is F _ _ _ _ _ _ _ in your cup”!
“M _ _ _ _ _ _ _-H _ _ _ _ good to the last drop”!
“A _ _ _ _ _ _ _ _-E _ _ _ _ _ _ _ don’t leave home without it”!
“_ _ _, what can brown do for you”?
“it pays to D _ _ _ _ _ _ _ r”!
“M _ _ _ _ _ _-C _ _ _, master the possibilities”!
“only from the mind of M _ _ _ _ _ _ a”!
“is it live or M _ _ _ _ _ _ x”?
“_ _, we bring good things to life”!
“how do you spell relief, _ _ _ _ _ _ _ s”?
“tum, tum, tum, tum, _ _ _s”!
“who could ask for anything more, T _ _ _ _ a”?
“P_ _ _ _, the choice of a new generation”?
“have a C _ _ _ and a smile”!

"N _ _ _, just do it"!
"T _ _ _ _, takes a lickin, keeps on tickin"!
"wait til we get our H _ _ _ _ on you"!
"G _ _ _ _ _ _ _ _, the thirst quencher"!
"N _ _ _ _ _, little, yellow, different"!
"the E _ _ _ _ _ _ _ _ _ _ bunny, it keeps going and going"!
"D _ _ _ _ _ _ _ _, the copper top battery"!
"built F _ _ _ tough"!
"S _ _ _ _ _ _ _ _, snuggly soft"!
"roll a R _ _ _ to your pal"!
"all temperature C _ _ _r"!
"C _ _ _ _, the heartbeat of America"!
"S _ _ _ _ _ _, a new kind of company, a new kind of car"!
"viva V _ _ _ _ _ _a"!
"C _ _ _ _ _ _, for when the time is right"!
"_ & _ Root beer's got that frosty mug taste"!
"B _ _ _ _ _ _ _ _ _ _, the king of beers"!
"it's M _ _ _ _ _ _ time"!
"M _ _ _ _ _ _ - L _ _ _, tastes great, less filling"!
"C _ _ _ _ _ _ _ _- M _ _ _ _ _ _, everybody's got a little captain in them"!
"ho, ho, ho, Green _ _ _ _t"!
"R _ _ _ a _ _ _ _, the San Francisco treat"!
"D _ _ _ _ _ 's, avoid the noid"!

"it's not delivery, it's D _ _ _ _ _ o"!
"how many licks to the center of T _ _ _ _ _ _- _ _ p"!
"don't squeeze the C _ _ _ _ _n"!
"D _ _ _ _ _ - D _ _ _ _ _, time to make the donuts"!
"K _ _ _ _ mac & cheese, it's the cheesiest"!
"F _ _ _ _ _ _ F _ _ _ _ _, they're great"!
"R _ _ _ _ _ B _ _ _, two scoops"!
"W _ _ _ _ _ _ _, the breakfast of champions"!
"L _ _ _ cereal, hey Mikey, he likes it'!
"Got _ _ _ k"?
"take the N _ _ _ _ _ plunge"!
"H _ _ _ _, H _ _ _ _, H _ _ _ _, wimpy, wimpy, wimpy"!
"leggo my _ _ _ o"!
"L _ _'s potato chips, bet you can't eat just one"!
"C _ _ _ _ _ _, they're the cheesiest"!
"P _ _ _ _ _ _ _, once you pop you can't stop"!
"_&_ s, melts in your mouth, not in your hands"!
"S _ _ _ _ _ _ _, taste the rainbow"!
"ooo la la, V _ _ _ _ S _ _ _ _ _ n"!
"H _ _ _ D _ _ _ _, you can do it, we can help"!
"L _ _ _'s, let's build something together"!
"P _ _ _ _ _ _ _ _ _, softens hands while you do the dishes"!

“get a little closer with A _ _ _ _ extra dry”!
“V _ _ _ _ _ _ _ _ intensive care, let the healing begin”
“brush your breath with D _ _ _ _ _ e”!
“S _ _ _ _ _ _ _ _ satisfies you”!
“B _ _ _ _ _, the quicker picker upper”!
“don’t get mad, get G _ _ d”!
“Im gonna get me some K_ _ _ _ _ _ _&_ _ _ s”!
“double your pleasure with D _ _ _ _ _ _ _ _ _ gum”!
“what would you do for a K _ _ _ _ _ _ _ _ bar”?
“C _ _ _ f _ _- a, eat more chikin”!
“W _ _ M _ _ _, save money, live better”!

As you can see, the agenda-pushing crowd has purposely and thoroughly “programmed” us with an expanded, yet surreal, vocabulary! Do not forget, acquitted and condemned by our choice of words, my brethren! All of the above jingles/slogans, in tandem with the manipu-list, have had an undeniable effect upon our psychology!

When we break down these two lists an interesting paradox emerges. The jingle/slogan list appeals directly to the mind that has been effectively blurred by the manipu-list and vice versa! Sex, gluttony, covetousness, usury, wants, promiscuity and drunkenness, to name a few, have been transformed to bear less consequence than Scripture merits! They have been effectively relabeled from sin to: "character traits" or "bad habits" or "vices". Word manipulation by, for and through commercialism "heaps a big ole pile of hurtin" against us!

In what measure, I'm not really sure, for I don't hold the scale in my hand, God does! Judgment day looms for all of us, inevitably, so either we continue to trudge down this path of perdition or change paths! That's it!

How? As I stated in the previous chapter, Christ! He is our only hope of restoration, sanctification, salvation and redemption! The spiritual war and battles are ever ongoing, ever with us and ever so powerful! So wake up, be aware of the enemy's tactics, grab the sword of Truth, put on the full armor of God and beg for mercy and strength! And in elementary fashion: "watch your mouth".

There is no way over, under, out of or around this war, there is only going through it, so arm you against the fiery, manipulative arrows to prepare for the days that lie ahead as time dwindles! Start with identifying the heavens for what they are, the heavens- not "space"!

Another area of manipulation and word bending you had better get a hold of is the music and video industry. This is a call to the mtv© generation. We all better pay very close attention to what we endorse, by way of the images and lyrics we allow in to our homes, thus our eyes and ears, through radio and videos! This would include cell phones, I-pods®, and mp3 players etc. Listen, would you open your front door and willingly invite satan himself into your home?

Well that is the ramification and result when we endorse the entertainment industries' drek, with their lyrical manipu-list. We are, in fact, swimming in a "culture of corruption"! Know the enemy, people! There is a four part series titled "They sold their souls for rock and roll" that every professing believer should acquire. There is also, on you tube®, a series called "hells bells2" that should be observed in its' entirety. They are startling, revealing and will change your minds towards a lot of the songs and bands we all grew up with. They'll change your mind about the whole music industry. Again, you must learn to recognize the enemy's seeds of discord or you will perish! Again, "**do not be partners**"!

You see, depending on your age, there is an entire generation of us who grew up with the technical boom. Any one born post nuclear should be able to identify the snowball effect of entertainments' grasp on the human mind. There was a time when merely chewing gum in school was a punishable offense. Now we have a gun wielding, pregnancy pact making, I want my respect gang-banger mentality rampid in our schools.

Cities, suburbs, black, white, rich, poor; it doesn't matter! The entertainment culture is an indiscriminate, equal opportunity, impressionable-child killer! What was culture shock to us is boring to them. Culture shock for them means even more outrageous thrill and shock than should ever be acceptable! Just refer back to chapter four and the pre-teens who find fun at the behest of injuring others-it's a thrill to their self-esteem deprived little thinking! Can you say flash mob?

These are our children, people! If all this technology were so stinking beneficial, it wouldn't be this way! As it is, we "sacrifice our children to Molech" by way of the modern day, anti-God, soothsaying mindset! Perhaps it's just me, but my mind cannot fathom Yahshua Jesus endorsing passing out condoms in school, can yours? Will they ever be able to recover? Will time allow it?

Stop asking "wwjd- what would Jesus do"? This is another deceptively clever manipulation of words! Wwjd? Why? Are you going to "drink the cup he drank"? You never again have to ask "wwjd"; he's already done all he's going to ever do physically! Wwjd? Die for you!!!!!!!!!!!!

Chapter 8: Time Passages: Creation Manipulation's Origins

Time! Ah, the fabric by which Yahweh has housed his creation! It escapes our grasp yet simultaneously marks out our meager existence. We can't touch it, see it, hear it, capture it or replace it, we can only observe its' passage. It precedes us, leads us and ultimately leaves us in its' wake! With every chip of each passing generation, time obediently marches on, and will until its' own God given destiny is realized!

A destiny we neither control nor really understand! We are encapsulated by it, mystified by it, perplexed by it and powerless to change it, as it too stands outside the bounds of our physical realm! As with scripture, music, words, numbers and electricity, time is of the spirit realm, of the unseen thus, of God! We don't know when it began and we don't know when it will end, although scripture prophesies and teaches its end is, from our viewpoint, perpetually close!

It is a fruitless endeavor to try to grapple time in to submission; and give it a good ole' man noogie. It doesn't work that way! We can't wrestle this one to our abusive liking, like has been allowed with the other spiritual endowments! We can't "Turn back time", "slow time down" or make "Time stand still" for "Time goes by"!

No, we are actually better at "Killing time" through "Time passages" and wasting it away than we are at utilizing it correctly! Time isn't "of the essence", time isn't "money", it is time, and certainly holds much more value than our currency would ever be able to purchase!

I know, I know; it's just a figure of speech, "time is money". Yeah, and words are nothing more than a sequence of letters right? Okay mister evolutionist, here's another test for you then; take everything scheduled, including the ability to schedule, and remove them and what remains? Do you see the pattern issuing forth again? It's the same as with every other spiritual gift bestowed upon man. If it were *actually* possible to remove them, nothing of our doing would remain!

It's frustrating isn't it? Oh how we love to go around touting our achievements as *god like*, yet if we could *really* subtract his doing, and were given but a blank canvas, our empty foolish futility would shine through with folly! We are oh so proud of ourselves aren't we? But before we tread too far down that particular path, you know, the one where I call you the worldly, atheist evolution spouting Nimrod, let's take a look at time, at least as far as we think we can understand it.

Let's begin in antediluvian antiquity; you got it, right back where it all started "In the beginning", the book of GENESIS! Let us first consider the time from the record of creation to the flood, from day one to the deluging end of that age. It is a time like no other in mans' history, full of what we would label as a legendary, mythical, mystical, magical and far-fetched time! It's the time that God's Word prophesied through men's names that redemption would be made available through God himself, remember? Brush up with a look at GENESIS chapters 1 – 6 to survey the landscape.

And as I've already stated numerous times, take scripture literally, even if for but a moment, and consider every word as factual, historical and literal! I know this is a daunting task for our thoroughly manipulated modern minds to fathom but do it! Consider it all real and take scripture, hence God, at his Word! Don't fiddle with nay saying, skepticism and doubt, open the narrow mind and learn the lessons contained within our past!

This was a really crazy, wicked, scary, corrupt, frightening, and violent time in mans' existence! People lived to be hundreds of years old, celestial beings inter-mingled with our race producing "men of renown", sons of God tweaked the earth, and began to train man to do the same, and giant Nephilim roamed the land.

Gigantic elaborate structures, including Stonehenge and the Great Pyramid were fashioned world wide, certain "secret" knowledge was extracted then hidden until the "times of the end", God was grieved at the heart and punishment was exacted!

This really was a weird time! So violently weird that Yahweh God said enough! It is also the time referenced by Christ, as descriptive of the end of days, the day of the Lord and the coming of Messiah in the clouds. Huh? What does that mean; the end will mirror the beginning? Is this to be two similar time period's book ending the whole of mankind?

Well, most only use the term violent in describing the similarities between pre-flood man and end times man, but I think there's much more to it than mere violence! In fact, scripture states wickedness, corruption *and* violence abounded! Why? Man was still rather fresh off of the choice of the tree in Eden, knowledge!

"As it was in the days of Noah, so it will be at the coming of the Son of Man". Christ spoke these words (MATTHEW 24: 37). And verse 38 lays out what *appears* to be mans' normal routines and daily activities, so let's begin with what verse 37 states. This is crucial to understanding where we are within the fabric of time and the reason by which an evolutionistic, psychologically polished mind-set must be abolished now, if it is not already too late!

In my own humble opinion, mans' fairly recent; paranormal activity, ghosts and spirits, area 51, "ufo", "uso" phenomenon, lights in the sky, grays, greens, contact, "abductions", jet propulsion labs, reverse engineering, SETI, NASA, asteral projection, remote viewing etc. is all linked directly to this time of antiquity. And I contend that, as it was back then, celestial and angelic mingling is the culprit, both good and bad! I do believe our whole modern existence, not limited to the etcetera list just compiled, forged by our creation manipulation hands to be the concluding chapters of what was begun way back when!

Why do I believe this? Well, first of all scripture prophesied it, read MATTHEW 24: 37 again. Second of all would be information and observation. That's right, available literature and available lenses placed in the front of my melon! Scripture charges us to observe and pay attention to the times we're in (MATTHEW 24: 42 and 25: 13), so that clears that one up. You may be wondering, though, what literature? Well it starts with scripture itself.

And then books like the "lost" books of Enoch and Jubilees, rediscovered in the late 1700's. Both shed some fascinating light on this antediluvian, pre-flood time period. And before you ask, *no* I **do not** know whether these "lost" books, or any of the discovered "lost" gospels should be a part of the biblical canon or not. What I do know is that there is a whole slew of information spelled out which peels back, completely, the cloak that the scientific community hopes modern man remains ignorant of.

In fact, I believe it to be quite possible that the entire canonized scripture could be *the* "sealed scroll" spoken of in scripture. It reads as such: "But you, Daniel, close up and seal the words of the scroll until the time of the end. Many will go here and there to increase knowledge" (DANIEL 12: 4).

I think DANIEL 12: 1-4 speaks of our "age", especially if the extra-biblical books were included! Think about it for a moment, before the "invention" of the **mass** production printing press in the 1800's, only higher-ups and church or assembly leaders had access to; read, study, research, dissect, translate and transliterate the Hebrew and Greek versions of the Word of God!

If the "lost" books are meant to be in the bible and man omitted them, God will deal with man appropriately, as is stated in REVELATION 22: 18-19! If they are rightly placed outside the canon of scripture, take solace knowing the bible itself references them. What I want to hone in on, however, is the information they, these "lost" books, convey; with my belief that they *may* possibly be part of the scroll "sealed" until the "end times", as well.

Bear in mind that the English translation of the KJV bible and the discovery of these "lost" books were withheld from man until this "age of enlightenment" dawned! Actually, it's possible that many "extra-biblical" manuscripts could be a part of the end time "sealed scroll", including the bible. This is conjecture on my behalf; I only hold this as possible, not fact.

However, I do hold this "age of enlightenment" invention, the mass producing printing press, to be a profound bridge to our modern, scientific, techno driven world, as it made God's Holy Word; with its' structured, encoded, gematria soaked knowledge, as well as these "lost" books of Enoch and Jubilees, available to the "dumb masses"! Not to mention the influx of mans' philosophies and hypothesizes, at that time, speculating whether or not scientifically calculable and linguistically decipherable knowledge is even hidden within the pages of scripture!

For instance: Isaac Newton, one of our aforementioned scientific pioneers, believed in and spent many a day searching for coded messages in the Holy bible. So did Ivan Panin and E.W. Bullinger, among others. Now in 2010 we know, by way of our computer programs, the bible to be encoded. I know there are those who claim bible codes don't exist but the evidence is pretty overwhelming and just as valid as their worldly scientific theories! Personally, I think Newton found what he was looking for, and opened up the way to today! A book titled 'Temple at the center of time' by the late David Flynn gives a remarkable peek in to the mindset of one Mr. Newton that you'll never be taught in school; because they are way too busy training us to be evolved monkeys!! Be that as it may,

The lost book of Enoch claims to be authored by righteous Enoch, yes the one plucked away as Elijah was, by Yahweh; and it tells of his direct interaction with celestial, angelic beings and outlines how man was given hidden knowledge from the spiritual realms! And I also consider it no coincidence that our explosion of tech**knowledge**y occurred only after scriptures' post printing press unveiling, coupled with the rediscovery of these "lost" books, as Jubilees, although detailed differently, echoes Enoch.

I also find it highly peculiar that there were mega structures sprinkled throughout the earths' landscape, I believe pre flood, bearing structural precision that matches solar, lunar and stellar dimensions. Both the "Great Pyramid" of Giza and Stonehenge, I contend, in agreement with others; were designed and built by celestial beings with the assistance of giant Nephillim, not man! Man may have been a part of a labor task force, but that's about it! These structures, whether angelic or demonic, bear witness to the creative Wisdom and knowledge of the Creator's creation, see ISAIAH 19: 19-20.

If the angels endowed man with this cosmic, stellar knowledge the Creator God sanctioned it. If demons and fallen angels were the culprit of this endowment: the knowledge was stolen then shown to man. A celestial fruit grab from knowledge's tree as well!

The other pyramids were, I believe, mans' attempt to replicate the "Great Pyramid", post flood! Stonehenge and the Great Pyramid, by *Bonnie Gaunt, is a revealing read for insight about this topic of celestial dimensions and how they apply both heavenly and within these structures. It is one of the reference materials outlined back in chapter five. *(Although I don't nescessarily agree with her conclusions about salvation, the mathematical evidence is overwhelming).

The unveiling of knowledge and timelines in the 1800's, found engraved within the stellar dimensioned, astral-knowledge bearing inner chambers of the "Great Pyramid", coupled with the information found in these "lost" books, stacked on top of the public availability of scripture, barely pre-ceded our modern "inventions" of: electricity, petroleum extraction, and flight, and machinery, computers, communications and this modern, man on the moon, "age of Aquarius"!

Also, it may surprise you to know that the "Great Pyramid" has been called the "Pillar of Enoch". Why? Could it be possible that the book of Enoch's manuscript, the supposed "ancient Egyptian technology", is what was found in that empty, upper chamber sarcophagus, instead of a pharaoh or his treasures? (This is just me wondering aloud).

This "Great Pyramid" strangely bears no hieroglyphics like its' man-made counterparts do, why? Because it was, I believe, both designed and built by celestial beings as both a "sign and a witness", with celestial measurements scribed by Enoch, then sealed until a later date. Or are all of these ancient temples being discovered, where human sacrifice on a worldwide scale, was taking place? I believe that's precisely what they were for. Could these ancient altars be no more than testaments to the seed of the serpent acting out his desire to both stain and eliminate the seed of the woman (DNA) so as to avoid the prophecy of GENESIS 3? That is what the whole angels mingling excursion was and again is, about. Only time will reveal the "great" pyramid's mystery. And like Stonehenge it was constructed pre-flood, thus the "**days of Noah**"!

Speaking of Noah's day, who taught man to ferment drink? Who taught man of dough and yeast, that God would at a later period in time, specifically command the Israelites on this issue? Who taught man to eat hallucinogenics or smoke mind-altering substances? Who taught man the summoning of the dead or the angelics; another issue God would address at a later time? Who revealed the significance of medicinal herbs and spices? Actually, all of these manipulations of God's creation were addressed at a later date in scripture, *through* the Israelites, in the Torah, *to* mankind.

The written Torah, containing God's commands to man, was necessary, it was not born in a vacuum. It was born out of necessity, as ElShaddi was dealing with areas where antediluvian man had trespassed, through angelic and demonic training, his divine laws. Man had to be, simultaneously, taught the error of early man's ways, and the standards by which Messiah would one day be accountable to the Word; all the while teaching the whole history of creation and man and redemption from Adam and Eve's disobedient act.

You see fellow man, scripture-God's Holy Word, deals with all issues of all time- past, present and future, simultaneously, all housed in one writing. And in multiple dimensions! Scripture deals with, and specifically contains information for, all creation! It's a time capsule with vast amounts of multi-dimensional laws, creeds, ordinances, precepts and commands. I believe the Holy Scriptures will one day be revealed to be hologram-ic in holographic nature, a visibly multi-dimensional writ. Not only is scripture the Word of God to all of mankind past, present and future, it's also a directive to all creation; God's Word began and sustains the act of creation! And "the Word became flesh", remember?

What I think everyone needs to be aware of, though, is this; the public availability of scripture via the mass producing printing press, the discovery of "lost books", the discovery of the blueprint knowledge housed in the sealed "Great Pyramid", the birth of our great nation, the harnessing of electrical currency and the extrapolation of combustible fossil fuels all occurred around the same time, during the age of mans' "enlightenment". Kabbala and alchemy, thus science, is built on this knowledge.

Not long after leaving the "dark ages" and right after exiting the "little ice age", as the scientists have labeled it, and after the meltdown of the North American ice sheet produced fertile soil to cultivate; man and agriculture combined to give birth to our modern, western hemisphere, civilization as we know it. Moreover, our great nation was first in many of mans' modern inventions. Three come immediately to mind:

electricity, the Eagle landing and the atomic bomb.

And then within the last couple centuries we find electricity mass produced upon growingly complex grid systems, while fossil fuels are extracted and mass produced, as structures of stone, brick and mortar give way to superstructures of metal, wood and glass; concrete jungles with high rise temples. Can you say tower of Babel? Here's a secret; Nimrod is still trying to build his tower.

Tunguska (Siberia 1908) preceded Hiroshima and Nagasaki by 3 decades. Celestial beings again, I believe, showing man the secret things of God's creative powers. Atomic bombs, hydrogen bombs-someone or some thing shared stellar knowledge with man again! Maybe Tunguska was a case of celestial beings showing man the result, before teaching him the details, of nuclear fission?

We have the flight of man with the invention of aircraft, moon landings, mars rover expeditions, "ufo/uso" sightings escalating, satellite activity rampant, space exploration expanding, quantum mechanics, quantum physics, nuclear fusion, nucleus and atom splitting, gene splicing, DNA cloning and the undeniable increase in knowledge, thus the "**end of days**"?

Can you discern the thread that binds these two ages of man together through times' passage? This weaving of past time with future time is the very same fruit threaded from Eden's' tree of knowledge down through the ages to us! What, specifically, is the binding thread? Creation manipulation, the ability to manipulate the resources Yahweh's awesome creation affords.

Could it be that this is what Christ spoke about when comparing these two time periods of man? Celestial beings disguised, this time, as "extra-terrestrial" or "alien intelligence" mingling and interacting with man? So much so that God will soon again exclaim, enough! If so, then scriptures' charge to know the signs of the times and "keep watch" must be adhered to! This is not a suggestion; it's a command.

Let's quickly compare our time with their time to see if our time is even remotely similar. Violence? Check! Wickedness? Check! Corruption? You guessed it, check! Trans-humanism? Check. Chimera and hybrid bioengineering? Check! Yeah, I'd surmise it is once again as the "**days of Noah**" were! Which, by default, places us in the "**end of days**"? I've been echoing scripture all along saying time is short! Do you understand yet?

Now, if you've ever read this book of Enoch, you know it spills the beans on a whole plethora of spiritual information. If you haven't, it identifies: angelic and demonic names, celestial responsibilities, heavenly portals, constellations, black holes, astrological calendars, atmospheric knowledge and it details the Kuiper belt, solar and lunar rotations and ecliptical cycles. It descriptively tells of how the fallen celestial beings taught antediluvian man "eternal secrets" he was not meant to learn, including astrology, astronomy and occultism!

It tells of the secrets of charms, enchantments, the color prism, written texts (Sumerian tablets?), music, stone cutting masonry, the extraction and use of precious stones, the extraction and elixir of precious metals, jewelry, make-up, agriculture, horticulture, weather patterns, geology, geometry, geography, directions (**news**), heaven, hell, war and its' weaponry, and blood-thirst! Basically, everything our "modern" world is.

Here's a teaser question for you, where is the square? It's one of the foundations of geometry right? Stonehenge is the circle; the Great Pyramid is the triangle, two more basic fundamentals of geometry, right? But where is that single undeniable four thousand year old, pre-historic, celestially designed, square stone structure that completes the hat trick? Ok, enough with the teaser, back to the "lost" manuscripts.

It, the book of Enoch, also contains prophecies about the coming Messiah; his resurrection and the redemption of fallen man, his return and eternal judgments! It is similar to the book of REVELATION in flavor and was known by the apostles, and even referenced in the book of JUDE (JUDE 1: 14)!

You might as well go ahead and read the entire book of JUDE; it only consists of one chapter and 24 verses, as it describes a dead ringer to evolution's trainings! It's the last general epistle before the book of REVELATION.

Moving on, I could have very well included this “book of Enoch” and even the “book of Jubilees”, with the reference material outlined in chapter five, as the writers of scripture were aware of them! There are other books that fit this category as well, but I leave that to you to find.

They are highly informative books of information, and my only reason for not including them in the study material mentioned in chapter 5, is apprehension! My apprehension lies in the possibility of demonic spirits being dangerously summoned through charms and enchantments by name, by us“dumb mass” common folk! And I do believe that these celestial names were invoked, by name, before the bolt kissed the kite and even more so, sadly, by the masses today.

Remember back in chapter six and the weight words carry? If you are compelled to get this book, please **do not** utter the names in it! Remember, this is a spiritual war and invoking spirits will only grieve the Holy Spirit and render you helpless and hopeless, so do not fall prey to curiosity! Come on, would man do that?

Yes! I do believe "enlightened" man is guilty of the same curiosity, invocation and false worship that led to Yahweh's' grievance and ultimate wiping out of antediluvian man! As I've stated numerous times, all facets of our physical existence find their origin in the spiritual realms! And that statement is in agreement with scripture.

And scripture declares them all to have originated from our Creator God; so bypassing God for celestial attention is trespassing! Bypassing God's all powerful Wisdom for celestial knowledge is a repeat of the day's of Noah, a replay of the garden of Eden *and* descriptive of modern man!

What I cannot possibly know, though, is which celestial beings, angelic or demonic, then turned and educated man with these spiritual elements! Electricity, words, numbers, science, music and yes-even time, all impact our realm from outside our ability to create or invent or fashion them! Therefore modern, evolution-thinking man plays the fool when claiming them as his own! Talk about opening Pandora's Box of creation manipulation!

I had often wondered what man did that was so egregious that YHWH would flood him out. God could have ended it all before it even began, back in the garden, but he didn't. He allowed mankind to exist until the flood, so what happened post Eden yet pre flood? Sorcery, summoning, mingling seed, invoking and witchcraft in God's name, gross idolatry and the manipulation of God's creation are my guess!

Yep, man invoked then worshipped these celestial beings as *gods*, learned from them, applied what they learned and bypassed the worship of the true Living God himself! Sound eerily familiar? It's a forfeiture of Wisdom in lieu of gaining knowledge. This forfeiture, I believe, broke our Heavenly Father's heart and struck him in the heart with justified anger, much like a child's disobedience breaks the parent's heart.

Can you now see why, in an earlier chapter, I labeled the tree of the knowledge of good and evil as such: the good knowledge? Scripture, the very Word of God! What's the evil, bad knowledge? We're living in it; twisted scripture! Do you see it yet? Do you discern it?

In this sophisticatedly modern, evolution enticed, psychologically drenched world, man continues to bite of the fruit of knowledge and bypass the Wisdom filled Word of, and worship of, God Most High! Now, to have lesser *gods* we need medium range entities don't we? Higher than us yet much lower than the Creator, right? Read JUDE verse 6 again. That is what the celestial beings were guilty of deceiving man with! They portrayed themselves as "*gods*" and then man worshipped them as such! This explains the very first and most important of the 10 Commandments! Make sense yet?

I surmise that the Egyptian, Assyrian, Sumerian, Medo-Persian, Phoenician, Celtic, Nordic, Babylonian, Grecian and Roman "*gods*" weren't myth and legend! They were antediluvian beings; imbedded in mans' memory, verbally passed on and again worshipped postdiluvian! Which is when stone and rock monoliths, pillars, altars, temples and statues began to be erected or in scripted to "the *gods*" with mans' hands. This explains the placement of the second commandment.

This is also when everything from trinkets to asherah poles, statues to zodiacs, and pictures to engraved images began to be carved or fashioned with mans' hands. Once again mans' propensity for idolatry reared its' ugly head! It was there pre flood; it was there post flood, only this time man was emulating their remembered celestial *gods* with what their *gods* trained them in. Do the tower of Babel and the name Nimrod, coupled with a "universal" language ring a bell now? This emulation is also precisely why, at a later time of his choosing, Yahweh pierced right through time and exclaimed here "**I AM THAT I AM**" to Moses and the Israelites in a pure, heavenly, Hebrew language!

There were angels with heavenly knowledge and powers that were worshipped as *gods*, hence the inundation of earth with water! And I hold it possible that all those mythical, legendary creatures of folk lure: dragons, unicorns, gargoyles, and the sphinx etc. were probably based on real antediluvian memories. And their deceptive wickedness, coupled with mans mingling, invoking and worship of them, through the Tetragrammaton – later dealt with through the third commandment - brought to conclusion that age of man! But what didn't perish with the flood? Time.

And that extinction by flood gave birth to the next linguistic age of man, beginning with the replenishing and repopulation of earth through Noah's bloodline! And that opened a whole *new* "can of worms", with the earth being divided simultaneously as mans' life expectancy was cut almost in half. Could this be a description of pre and post Pangaea?

Stop reading, pick up your bible and *lay a sheet down* with GENESIS 10: 5 overlaying GENESIS 11: 18, with Peleg being the commonality. I know the scientists claim everything, under carbon dating, to be millions and billions of years old but as I said earlier, I don't fully trust man and his science since their motives aren't really known, I trust God. With man we have the potential for error, not so with God. Ok, now that you're back, what other popcorn trail has time left on its' path?

The earliest one-world government was dismantled, a common language was confused, and man was disbursed abroad, which man somehow knew would happen, and a monumental shrine was left half constructed. The Hebrew Abraham was chosen, Sodom and Gomorrah were buried and made an example of,

Jacob became the house of Israel, Joseph was sold then promoted and Israel became enslaved in Egypt. Bringing an end to the patriarchal age.

Plagues were poured out, not to smite the Egyptians only but to smite the Egyptian "gods" as well. Passover was instituted, the sea was parted, and Torah was given, as the Hebrew language was bestowed, the wilderness was wandered, the sacrificial system was defined, the ark was fashioned and the temple was temporarily tabernacled!

The Jordan was parted, the Promised Land entered, Jericho fell, wars were fought, judges were appointed, the kingdom was established through David, the temple was built by Solomon and God spoke to man through the prophets. The ten northern tribes were divorced, scattered then "lost", Judah was carried away to Babylon and a remnant returned. Bringing an end to the Israelite kingdom age. And what persisted? Time.

The minor prophets spoke, the Creator became the created, the kingdom of heaven was preached, the created killed the Creator, the sacrificial system was given sabbatical, the Torah spring feasts fulfilled

(Passover/unleavened bread), the Creator resurrected (first fruits), the Holy Spirit poured forth (Shavuot), the church age born and the "times of the gentiles" dawned!

Time didn't end with the patriarchs, the giving of Torah, the judges, the prophets, the gospel, and the epistles of Paul, the general epistles or the REVELATION of Christ to JOHN! And time as we know it, won't end until Yahweh declares it so, with the fulfillment of the fall feasts and the return of Messiah! Meanwhile, Yahshua Jesus told all of mankind what to look for as the end of this age approaches: the replay of a pre-flood mind-set, a return to summoning spirits, the return of "channeling", divining and spiritualism and the pride lifting audacity of creation manipulation! Just "As it was in the days of Noah"!

Look for the heart bent toward the idolatrous, false worship of: celestial beings, knowledge, earth, nature and now because of evolution, add man to this list! Man is again guilty of worshipping everything under the sun; accept God!

We are still guilty of worshiping everything in and under the heavens accept the one who placed it all there! And what does that define? What do you know: I, my, me, mine modern man. Just as in the days of Noah, just as scripture prophesies, Yahweh God is overlooked, forgotten and snubbed by man! Horoscopes are given daily ritualistic attention while God is, so we've been trained to believe, non-existent! Oh the peril we face!!!

Only this time, the wrath poured forth will be eternal! Fire and brimstone, plagues, famine, disease, the shaking of the foundations of heaven and earth themselves, culminating with the eternal establishment of YHWH's heavenly kingdom on earth! "Your kingdom come, your will be done on earth as it is in heaven" (MATTHEW 6: 10).

You see fellow man, with the passing of time, scripture hasn't become "outdated", as the evolution pushers would hope for you to believe, it is being fulfilled! Man can turn his back on God all he wants and doubt the validity of scriptures' God-breathed Word all he wants, but scripture will not be thwarted by such feeble attempts and man does so to his own peril.

The truth is his Word is Truth! Yes, absolute, concrete, dependable and time tested

Truth! *Lay a sheet down.* Historically verifiable! *Lay a sheet down.* Archaeologically substantiated! *Lay a sheet down.* Scientifically denied, *lay a sheet down.* Scientifically verified, *lay a sheet down.* Coming to pass, *lay a sheet down.* And almost to full fruition; *lay down the final sheet*!

Can you see now why it is so imperative to get off of this track of futility? Does your heart discern the urgent necessity to reject the putrid "**theory** of evolution" outright? Can we, man, finally begin to comprehend the damage inflicted by this diabolical training yet? Are our hearts able to understand the wretchedness the purveyors of evolution have saturated us with?

Through adversarial words, programming, thorough training, entertainment and education, man has attempted to train us to be "monkey-minded"; that we "evolved" from primates, that we are merely the "descendants" of apes; that we are but chimps that somewhere along the line "stood upright", that we inherited no created features of our Creator; that we are "dumb mass" monkeys and that "God does not exist"!!!

"In a nutshell", "bottom line", "the whole kit and caboodle" states, emphatically, that we are not created; that there is no "intelligent design" behind it all and that we are but the result of some cosmic, zero origin "big bang"! And on top of that don't you dare question their point of view! Never-mind the lack of evidence; this is the "age of enlightenment" remember? I say <u>sarcastically</u>, we're enlightened people so just sit down, shut up and keep your "dumb mass opinion" to yourselves, you "Dumb Mass monkeys…"

Umm, no! I won't "stifle it" any longer, time won't allow me! I've had it with the court backed strong-arm tactics of the progressively agnostic, boldly atheistic, psychology smothered, evolutionary trainings of modern man! It's **corrupt**, it's **wicked** and it produces **violence**! No, this is my response to it all!

I'm not quitting the fight this time, as I did during teenager dom! No, I will lay science at the foot of the staked cross, the one mans' sin was sacrificially nailed to, as the Wisdom filled attribute of All Mighty El Shaddi that it truly is! And I refuse to shrink back again!

It is high time man wakes up or pays the heaviest of prices, deserved eternal separation from his Creator; which happens to be the ultimate "goal" of the serpent: beginning with its conversation in the garden, at the tree, with Eve, about the Creator! Time is short, the scroll is unrolling and Messiah's return is imminent, so be aware and begin to prepare!!

Now, what else do we know about time? What does scripture say about time? What does falsehood vs. Truth say about time? Well, the **Word states (teaches)** we are in the "times of the gentiles", which is temporary, while the **world speculates (trains)** the Mayan calendar and Nostrodamus quatrains!

The **Word states** the "governing lights" in the firmament are to serve as "signs to mark the seasons" and "to separate the light from the darkness"; while the **world speculates** everything is indifferent, with no inherent design or reason! The **Word states** time is finite, measured and at our Creators' disposal, while the **world speculates** time can be traveled via "wormholes"!

The one thing that should jump out at all of us is that the **Word states (teaches)** while the **world speculates (trains)**! Thus is the folly of man! Please, I beg you fellow man, spend time with your noses buried in the books of Wisdom!

They are, again; JOB, PSALMS, PROVERBS, ECCLESIASTES and SONGS OF SOLOMON. The vital distinction between what the **world** speculates vs. What the **Word** states is life and death crucial as time unwinds!

Remember 1st CORINTHIANS 3: 19, which **states, "**The wisdom of this **world** is foolishness in God's sight"! And JAMES 4: 4 **states** "You adulterous people, don't you know that friendship with the **world** is hatred toward God"? And again PSALMS 9: 8 **states** "He will judge the **world**"? And once more in PSALMS 96: 13 "He will judge the **world**"?

The **Word states** while the **world speculates**! The **Word teaches** whereas the **world trains**. We're in the danger zone here people! Our view of the Word is skewed, our view of the world is skewed and our view of time, through our modernization, is severely skewed! Our "fast food", "24 hours a day", "instant gratification", "got to have it now", "no time to wait", "rapid response", "time is of the essence", "time is money" attitudes only expedites sheol's arrival!

From the "womb to the tomb", from "birth back to earth", from the "cradle to the grave" is our lot in life so get over it Mr. modern evolution man! People, either rip the intravenous drip of evolutions theory from our thinking or r.i.p! Rip or R.I.P. it's your "free willed" choice! But make it quickly for time is running out on all of us! And I'm not an alarmist I'm a realist!

I once was an alarmist and got caught up in the guessing game of predictions, dates and future telling, as pertaining to the **Word** not the **world**, when something miraculous happened to me; I got diagnosed with a very rare abdominal tumor. It was an excruciating experience to endure but praise Yahweh God, my wife and my family has been a rock through it all! And on the other side of this experience, looking back, I call it all a blessing!

While recovering I turned, as always, to scripture, and found that science isn't opposed to God, as I have erroneously been taught, it is an attribute, aspect, quality and partial slice of God's sovereign Wisdom! A mere trait of our Holy, pure, perfect, all knowing Creator! "Our God is an awesome God" to be sure!

Listen fellow man, we were never given the right to manipulate his creation, we disobeyed and stole it remember? And our means of doing this, science, is stolen as well! Antediluvian man and his ancestral offspring, modern man, yes you and me, are both guilty of a burning propensity to manipulate our Creators creation unto servitude, yet we demand our "r-e-s-p-e-c-t", ha!

Science is but a portion of his magnificent, timeless Wisdom! Our fallen nature with its' misguided ideology, skewed view of gregorian time, error ridden view of technology and misappropriated view of wisdom can't taint Holy Wisdom as "the fear of the Lord is the beginning of Wisdom, and knowledge of the Holy One is understanding" (PROVERBS 9: 10). Time will stand as long as he deems necessary, according to his good will and Wisdom! Amen! We'll not grapple this one into submission. Nope, that good ole man noogie will have to wait for another time- pun intended.

Time, speech, calculation, Wisdom, scripture and creation itself are all of the Holy Spirit, unseen God! Not unknown, unseen! There is a big difference! Antediluvian man wasn't spared: how arrogant of modern man to think that he will be spared and able to claim "we didn't know"! Hogwash, hiding behind "ignorance is bliss" won't work with God!

Okay, what of the square? I know in "Stonehenge and the Great Pyramid" by *Bonnie Gaunt, Ezekiel's vision of the temple at Jerusalem, and its' heavenly dimensions are outlined and contrasted with Stonehenge and the Great Pyramid. But Ezekiel's temple contained an altar for sacrifice and offerings (EZEKIEL 43: 18-26), and this was the foreshadowing system fulfilled through Christ's impalement.

Read on through EZEKIEL 44: 1-3, and take note of the sealed East Gate, still sealed to this day, which is the gate passed through by Messiah riding on a mule, being hailed as "Hosanna!" (JOHN 12: 4) and (LUKE 19: 37-38). The footprint or "blueprint" of that temple (called Herod's temple) remains but both Solomon's and Herod's temples were built post flood.

I'm looking for that 4,000-year-old structure that completes the antediluvian tri-fecta of geometrical design: triangle (Pyramid), circle (Stonehenge) and square. Which leads me to the book of REVELATION (21: 10-26), the description of a heavenly dimensioned, perfect square. Could it be? This future kingdom temple is as the two other, antediluvian, structures: spiritually designed with celestial stellar dimensions.

Only this one will be, we know, of God the Father, whereas the other two are of spiritual design, but from which side of the spirit realm we don't really know! And the intriguing factor to me is its' square (cube) dimensions, the trilogy of geometry 101 completed? Only *time* will tell!

So, mister worldly, atheist evolution spouting, Babel building Nimrod: with all your liths and gliths out of the way, what now? **Who** will you blame when the clock strikes midnight? **What** will you do when time's up? **Where** will you turn when the last grain of sand falls? **When** will you realize the end is nearer than you think? **And why**, with all of your degrees and diplomas, does your heart still lick at the altar of Darwin's theory? And finally, **how** will you handle it when the last trump blows? Just scientifically deduce for me an answer to these time honored questions. And hurry for the singularity fast approaches!

"Oooh oooh aaah aaah na-noo na-noo beep oooh aaah? Will the chimps figure out who we are, thus who they are?"

Chapter 9: Speculation Proclamation

It seems to me, as I proclaim my own chapter of speculation, that we've meandered our way to the completion of our journey through the minefield of evolutions' psychology laced trainings. Right from the very start I said it would be harsh, and that it is! But it's fair, clean and to the point! I have purposely kept it clean and free of words that curse and slander!

I have named only a few names, limited to the scientific minds of Newton, Einstein, Galileo, Copernicus and one Mr. Hawking, as to avoid being slanderous! Ok yeah, I might have mentioned Freud, Pavlov and Darwin once or twice but throughout I've slandered the theory not the individual man!

I have done the same with the new-age self-help "gurus" *Hubbard and Robbins, but I've been careful to slay the trainings rather than slander the man. Ok, I've named some band names as well; as far as deciphering the meanings of the band names or reciting the "artistic" lyrics of the music industry, they pretty much speak for themselves!

I've purposely littered this journey with many clichés, merely to contrast the **world**s' temporal language with the **Word**s eternal language. I've tried to hold the "have your cake and eat it too" jargon-istic babble at bay! I've tried to equally point the finger, both individually and collectively, as we all share the shameful guilt and bear the deserved consequences! Individually and as a whole we are still fallen and our culpability remains, therefore, still Adam and Eve-mankind!

I've tried to avoid any interpretation of scripture, as that is Holy Spirit's duty not mine! In truth, I don't believe man is to interpret scripture; he is to divide it correctly! There is a difference!

I've quoted some scripture and left others for you to seek out on your own, which is my shrewd attempt to get us all to be like the Bereans as stated in ACTS 17: 11. Be a scripture examiner, the cross is ours to bear! Need proof? Here's your verification, read MATTHEW 10: 38 and 16: 24, MARK 8: 34 and LUKE 9: 23 and 14: 27 before continuing.

Do you see the intriguing revelation in these words Christ speaks? They were spoken before his impalement; he knew how he was to meet death! But how are we to follow or imitate this? Are we to go build some gallows and paste ourselves to them? That would be the physical aspect to his words but remember Messiah was the Word incarnate so enter again the double-edged, bone and marrow-separating sword! Oh he meant it literally, but physically or spiritually? Or both?

Well what, outside of prophecy being fulfilled in obedience to the one who authored it, would possess an innocent Jewish man to allow himself to be beaten, demoralized, persecuted, scourged then spiked to a cross without once complaining? He was possessed! But not a demonic possession, no, a Holy Spirit possession! He is the fulfillment of the Word!

What must be forfeited for this possession to occur? Free will, the new-age movement guru's chief talking point! That's the cross we bear, the laying down of our free will and the picking up of Yahweh's will! Messiah came not only to reveal the Father and his coming kingdom, he also came to show that a human being could actually fully obey God's commands, thus fulfilling the Torah; and the only way for man to do this is being filled with the power of the Holy Spirit! How does one acquire the filling of the Holy Spirit? Faith in the death and resurrection of Christ and the suspension of free will: "Yet not as I will, but as you will" (MATTHEW 26: 39). And this he suspended unto death. So much for goal setting workshops!

How? Read ROMANS 12: 1-2. And this is not an interpretation; it's merely dividing the Word at face value; if he says it he means it, literally! There are no "shades of gray" with God as there are with manipulatively evolutionized man.

It all boils down to be faith in God and his Word; the very Word evolution seeks to teach away. That's how Christ fulfilled prophecy and proclaimed - "It is finished"! Faithfully obey or faithfully repent! Obey Torah, which scripture reveals as everlasting, or repent at the blood of the cross-, which scripture defines as Life! Not Jewish? Doesn't matter, neither were Abraham and Sarah. Remember in chapter three we saw how our identities are being methodically eroded by our "creations"?

Nationalities are important, but scripture is the true story of God, his creation, his dealing with the redemption of man as a whole, and then it becomes about individual peoples. Adam and Eve did not yet have a nationality in the garden either, Remember? They had identities but not a **nation**ality to call their own after the fall! Let's expound on this a little bit.

This should remove the necessity to ponder our origins ever again! If, however, evolution has such a hold on you that you can't let it go, I pray earnestly that you pray about it. Pray as to avoid becoming "the prey"; pick teaching over training.

In chapter three we verified through scripture that Abram and Sarai, once YaHWeH breathed a ה into them, became Abraham and Sarah, the father and mother of nations-plural. But what about Abraham and Hagar the Egyptian, who did they become? Well they also bore a great nation, a people, (GENESIS 17: 20). Well, we have right here in the persons of this one man and these two women the origins of Judaism, Christianity and Islam.

These three people are the beginning point of many peoples and nations (including America and Britain) and they are also symbolic of both physical and spiritual Life (GALATIANS 4: 22-23), Isaac representing the spiritual and Ishmael representing the physical. But keep tracing back and you wind up with Noah, his sons and their wives. And continue tracing back through Noah, through Enoch and through Seth, and we wind up back in the garden with our original earthly mommy and daddy-Adam and Eve! Not Adam and Steve, not Amanda and Eve, Adam and Eve!

What's the importance? It is our individual and collective identities within the veracity of life that bring importance. Mans' free willed nature, again which contains the inclusion of woman, is still spiritually "fallen" and we "can't get up"! Through effective psychological seedlings comes effective evolution trainings and through effective mind manipulation comes mankind's loosey-goosey liberal theology; as a society we endorse redefining marriage, slaughtering the unborn, looking the other way and eliminating our origins altogether.

It's our free willed, disobedient human nature (Ishmael) that hinders, hence Christ on the cross (Isaac)! Hence Messiah on Mount Moriah! Do not forget that God himself pierced time and became man to rectify what began with Adam and Eve in the garden.

We, yes all of mankind, were there at that moment the fruit was bitten. What? Yes we are all culpable, guilty, bear the iniquity of and carry the condemnation of that original sin! How? Genetics. Our DNA, cellular and blood-type heritage bear our guilt. Huh? We are man! Not a man, man, as in mankind! It is important to remember that when YHWH looks upon us he peers *through* time not within time, as is our lot in life. The earth was cursed, therefore the whole of mankind, past, present and future, as our physical body is formed from the dust of the earth.

This guilt we bear and share is played out on the infinitesimal, cellular level. It's inherited through our blood. See the bigger picture of the sacrificial shedding of blood, first of animals then of God himself, necessary to cover the guilt of sin. And that blood running through your veins right now is descended form the first man and woman through replication, replenishment and inheritance.

What do I mean? Well, scripture is the entire record of creations' allotted time and

God's dealings with it. Look up; are the heavens still there? Yes. Look around; is man still here? Yes. From Gods' point of view, outside the bounds of time, everything is already accomplished. It was the moment "In the beginning" was declared. He sees the beginning to the end, the end from the beginning, all at once. There is no "age" of God! He is **from** Everlasting **to** Everlasting, got it?

We, however, see only as time allows by its' unfolding. So when God created man in his image, it includes those not even born yet! When God formed man from the dust of the earth, it includes those not even born yet! As a whole, we are mankind: collectively, then individually. And we are all guilty without the atoning blood of Christ, individually then collectively. And know this, those who point the finger at a particular group and claim, "they crucified Messiah", are in error! The Romans blame the Jews, the Jews blame the Romans and on and on we go. Here's the plain and simple truth; every single person who has ever lived, and every person that will ever live, are to blame. We are all to blame collectively, for his sacrifice is for us all, individually!

If you feel as though you bear no blame, then you feel as though you need no redemption and if that be the case, then you nullify the blood of the cross for yourself. Don't do this! Instead, man up, be honest and humble, count your accountability, confess your guilt and be saved. Then breathe.

When God breathed the breath of life into mans' nostrils (GENESIS 2: 7), the means by which human life is sustained was born. It is the very means by which those not yet even born will be sustained as well. Look around; is breathing still required for mankind to survive? Yes!

Until Messiah comes and establishes his heavenly kingdom on earth, day six in creations' archives is still unfolding. Say what? Day six, in GENESIS 1: 26 is, from our point of view still happening, and will until time has completed its' course. Confusing? We are from below; Christ is from above. There is a distinction between the two made by the Creator of the two. Those below belong to the rules and regulations of the below; Blood, flesh and bone, and all of which that encompasses, namely Adam and Eve.

There's more. Day one is still ongoing as well. Huh? So is "In the beginning". Huh? Again, is time still with us? Yes. Are the heavens still above us? Yes. Is man still here? Yes. That is precisely how *we were there* in the garden when the forbidden fruit was eaten; we were there, at the tree, when sins were nailed and the blood was shed! Comprehend?

Picture the bible alive and open before you and remember two things 1) scripture is multi-layered and multi-dimensional, so we cannot use *only* our physical senses to draw from it, for the physical realm has its' physical limitations. And 2) ancient scripture is really a scroll, or a collection of scrolls, not only a book; and in its' original form it's not broken down into chapters and verses like our modern equivalent, as a scroll rolls upon itself.

We have to unroll the scroll completely from end to end. Or rip, symbolically of course, every page out of the bible and methodically lay them consecutively next to each other, and form a continuous writing from GENESIS to REVELATION.

Now, with that picture in your melon, take the first words in the first chapter of the first book, GENESIS 1; 1 "In the beginning" and picture that as page one. Now fasten the edges of that page in place. Picture two loops at the top and bottom of the left edges of this first page and hook them onto two fastened pegs positioned appropriately.

Now pull and stretch that all the way to where the book of REVELATION would be after laying the pages consecutively next to each other.

Keep stretching as you "time travel" to REVELATION 22: 21 "the grace of the Lord Jesus be with God's people. Amen", and hook the two loops at the top and bottom of the right edges of this page to its corresponding pegs. What you should have if picturing it correctly is GENESIS to REVELATION, the whole of scripture, stretched out before you with GENESIS 1: 1 "In the beginning God created" stretching across the whole entirety of the Word from GENESIS 1: 1 to REVELATION 22: 21.

Now take GENESIS 1: 3 "And God said let there be light', and there was light". Hook stretch then hook. Um, if your doubt needs assistance look at creation, is there still light? Yes, so day one is still happening too, as far as our understanding of time continuum goes. This hook, stretch and hook can be done with the whole creation record from GENESIS 1: 1 to GENESIS 1: 31, with each act of creation being stretched from GENESIS 1: 1 to REVELATION 22: 21. Technically, "In the beginning God created the heavens and the earth" is all encompassing.

This hook, stretch and hook can also be **cautiously** done with the book of REVELATION in the reverse order. And this one book, the book of REVELATION, I think more than any other, causes so much confusion to man because he tries to dissect and interpret it solely in physical terms! Meanwhile the spiritual aspect is reduced to only "*figurative and allegorical symbolism*". Stop it! Divide the Word correctly! Stop attempting to read and decipher scripture chronologically.

Has it ever occurred to you mister evolution drenched man, that the book of REVELATION is a, drum roll please, revelation! That it's a reveal, describing the spiritual dimension, which occurs simultaneously with our physical dimension? Perhaps it is much more than merely a book of "symbolic" and "figurative" metaphors, which is presently "the school of thought" for both the believing and the unbelieving? Here's an outdated novel idea, instead of interpreting it figuratively or symbolically, believe and obey it literally!

Listen, I'm not saying that scripture is without allegory and symbolism. I'm only indicating that the knee jerk reaction of always approaching it this way first, diminishes the possibility of its' literal implications. For example:

Let's take a literal look at the notorious four horsemen found in REVELATION 6: 2-8, bearing in mind that the present teaching in the church is that from REVELATION chapter four on is only futuristic prophecy:

Conquest, war, famine and death! As I confessed in the last chapter, I was once guilty of needing to know times, dates and the future so I would know; I convinced myself, what to do to prepare for the dark days that lie ahead. I was in error when I pre-supposed a future arrival of these prophesies. Maybe they are not exclusively bound for a future time. Perhaps they will increase and intensify as time unfolds.

Dark days are still on their way but my need to know who, what, where, when and why these four horsemen are, has subsided a bit! I don't know when they'll come but I do know that conquest, war, famine and death have been with mankind as far back as the antediluvian age. Yes, literally! Scripture may contain some symbolism but it is not figurative, it's spiritual. Literally! Hook, stretch and hook. Another chapter worthy of note is the infamous chapter 12, which reads more like a snapshot of events than only a prediction of them, hook, stretch and hook.

Ok, back to hooking GENESIS 1: 1 to REVELATION 22: 21. Everything created by our Creator is still in their created mode, therefore still happening. Now do you see it? The heavens, the earth, its' atmosphere, the lights in the expanse, days and nights and seasons, the seas and rivers and lakes of water, the land masses and the trees and vegetation are all still firmly established!

The birds of the air, creatures of the sea, the creatures and animals and livestock of the land are all still alive and present, therefore still happening. And last but most, man is still above the other blood and breath creations by way of speech, calculation, reason and his God-given, God-stamped image, got it?

The record of creation is, physically, still being recorded because this portion of creations' time is not yet finished, from mans' point of view. In God's eyes it's one snap shot of completion, for us time is still unfolding. Do you now see how we are Adam? We are Adam and Eve; we are mankind!

Praise Christ and what he accomplished on that cross because without God's redemption we would remain in a perpetual state of fallen Adam, fallen man. And we would never know of the glory of our Creator-God-El Shaddi-Yahweh, as the shed blood of an animal can never atone for the sin of man.

Praise Yahshua Elohim! Now, do our individually nationalized eroding identities really matter? Yes, but not nearly as much as his shed blood, sacrifice, salvation and reconciliation! He is the Aleph and Tav, Alpha and Omega, the First and the Last; who knows the End from the Beginning! **He is the Word** and **he is raised**! If you are a true believer that *is* your "born again", “new man in Christ”, “blessed” through Abraham identity, “any questions”?

In truth, the denying of him heaps condemnation and contributes to the success of our eroding identities! Can your mind fathom it mister anti-God atheist, Mr. Anti-Christ evolutionist? That’s right; you house the spirit of anti-Christ with such thinking! And denying the only road we have ever been or ever will be offered is an outright blaspheming of the Holy Spirit.

We are Adam! I know it’s boringly repetitive but it’s imperative to understand! We begin at birth in our fallen state and from that point the erosion ensues. No matter how much modern man elevates his notions of himself, it doesn’t end on a good note. Not in light of scriptures’ declaration.

If you feel I’m being judgmental know this; he who denies the Christ “stands condemned already” (JOHN 3: 18). I’m not judging you, as you already stand judged! It’s quite amazing how man loves to spout JOHN 3: 16 “and God so loved the world”, yet completely ignores a mere two verses later.

So if you claim the name, resolve within yourself to claim it even if the cost is your very life! That is what scripture teaches, and each apostle died for it. For the last time, wwjd? The question is already answered, die instead of sin: which scripture defines as trespassing Torah! He would certainly never ever deny God the Father like evolution does! If you truly are a believer, you are just simply not allowed to believe even the smallest, minute detail of evolutions' theoretical based training! You are not allowed to "have your cake and eat it too" as that is exactly what liberal theology allows.

When you truly are a believer, you will be at odds with much of today's secular, evolutionistic thinking! You cannot believe both, period! That only produces double-mindedness and an erroneous view of Messiah, as Christ declares, "Wisdom is proved right by her actions" (MATTHEW 11: 19-20). So the question still remains; which do you want, Yahweh's Wisdom or mans temporal knowledge? What are you, wheat or chaff? Have you been taught or trained?

For those who are admittedly non-believers, you are already identified and your identity cannot be eroded any further than that: although the days that are coming upon us all will leave you mystified, terrified and without hope. All you can do is try to drag as many down with you as you can, hence evolution. There is a better way; there is hope for you through Christ! If denial is your choice of free will, then God has indeed given you over to the inclinations of your heart! Moving on,

There is one aspect of the creation record that is housed beyond our comprehension and that's day seven, the Holy and Blessed day of our Creator's rest! Creation was completed so our Creator rested. That is the information given about the seventh day, which has no evening and no morning! The surface text declares this particular day to be purely spiritual and outside the bounds of time and outside our ability to understand or enter into it physically. This makes it scientifically immeasurable! Let the theories start flying!

But! Scripture does tell us in the New Testament that Christ became the seventh days' link to the Sabbath's rest. This is an abode outside the limits of time, outside the limits of physical understanding and speech and calculations, only available in the spirit and commanded to be observed in its' microcosm version of the weekly Sabbath!

A command still in effect today, regardless of what man thinks! Go to church on Sunday every week but know this; Christ never came to abolish the Torah law, he came and fulfilled it! He and his disciples observed the Sabbath; did you know that? Christ and his followers observed every Friday at sundown to Saturday at sundown as Sabbath. So shouldn't we, as his followers?

When dividing the Word correctly it includes the whole of scripture, not merely New Testament theology! To separate the two is a gross trespass of the Word. Liberal theologians have twisted the Word to soothe their own "itchy ears" (2nd TIMOTHY 4: 2-4). Like it or not man, Messiah and his disciples observed the weekly Sabbath, so shouldn't his true followers observe it as well? I'd like you to read GALATIANS 4: 24 and pay very close attention to something. Paul, scolded the churches in Galatia for desiring to come "under the law again". In verse 24 Paul points out the difference between the above-mentioned two Covenants (Isaac and Ishmael).

My brethren, both Passover and observance of the seventh day Sabbath preceded the giving of Torah on Mt. Sinai. The Passover was first instituted in EXODUS 12: 21-24. The Sabbath was first instituted in EXODUS 16: 23-30. These two Everlasting ordinances by the God of Israel were commanded as Everlasting ordinances long before Moses was called to ascend Mt Sinai to receive Torah Law, so divide the Word correctly! Moving on,

The biblical account of creation, which is the only account worth merit, speaks on a timetable too big and vast for our ability to "get it"! Explain this to me: "And there was evening, and there was morning- the first day". Each successive day of creation ends like this, "And there was evening and there was morning". But take note of the order in which God determines daily cycles. Evening comes first; foreshadowing the lunar/solar calendar the Israelites were commanded by God to observe, then comes morning.

My need for an explanation lies in the fact that our modern calendars are solar while God's remains solar/lunar. Why? And what are, or will be, the ramifications of this? It is as though we purposely do it contrary to his commands, and it has severely skewed our notions of time. 2nd PETER chapter three should be read and applied by all who claim to be of the body of Christ! Actually, all of scripture should be read and applied!

Another explanation needed is literary style. Hebrew is written and read right to left whereas we read and write left to right. Again, why? Is this just an ancient, sophisticatedly subtle stumbling block? Or is it full of a deeper truth?

I've heard it explained that those who are east of Jerusalem read right to left and those who are west of Jerusalem read left to right. I don't know the validity of this notion but the implications are pretty profound, don't you think? It implies both eastern and western hemispheres point to and are lynch-pinned by, at, to and through Jerusalem. Interesting huh? Scripture anyone?

Wrap this around your skull Mr. evolution man! Mt. Moriah, where Abraham was willing to sacrifice his son Isaac, and Mt. Zion, where Yahweh was willing to sacrifice his son Yahshua Jesus, are one and the same. *Lay down a sheet.* See GENESIS 22: 1-14. And where is this specific mountain located? You got it, Jerusalem!

There is something about this lynchpin city on this mountain that transcends our human faculties! It is the city and mountain of God, as chosen by God! Just go ask one of the modern religious pillars of Judaism, Christianity or Islam about the importance of this region, since they've been grappling over this one strip of land for centuries. Make that millennia; not only since 1948!

As the world quibbles and squabbles over Nostradamus and the Mayan calendar and solar flares and galactic alignments and the end of the world, I'll keep my eyes on Jerusalem, as that is a better indicator of what lies just beyond the horizon. I look to Jerusalem for indicators of the end times. When man divides that land, know the times we're in.

It is the "Promised Land", blessed by God, blessed of God and promised to Abraham by way of Isaac and Jacob. Jacob, thus Israel, inherits it. The house of Israel, not just the remnants of the tribe of Judah, which currently resides there, and those sanctified by the blood of Christ, inherits it. Read the Torah, the scroll of God, for verification of whom the promise was to. Whether you read right to left or left to right the promise remains the same.

As the new-agers, one-worlders, UN supporters, and Babel seeking Nimroders, with their progressive machine devouring the whole earth, inch ever closer to their ultimate universal one world goal, keep your eyes on Jerusalem. This city is the flashpoint, the real "ground zero"!

Listen, I don't mean to be condescending, and I know it may be difficult to decipher at times whether I'm targeting atheists or believers. But I come from the point of view which scripture declares, and I trust we each have enough conviction to know which side we each, individually, reside on! Just know this fellow man: there is a reason Islam seeks to destroy "Israel" and the "west", and supplant our Torah Law republics with their Sharia law dictatorships; SCRIPTURE!!!!

Since that day in May of 1948 when Judah was reborn as the nation of Israel in the Promised Land, the cry out of the Middle East is "destroy the west" and "wipe Israel off the map"! Do you know why? Why are America and

England the only supporters of that little strip of land, while the rest of the world hates Judah inhabited Israel? SCRIPTURE!!! We are part of the whole "House of Jacob" (which even includes we gentiles) that's why! In a most elementary sibling rivalry, Sarah and Hagar's offspring (Isaac and Ishmael) are still duking it out over this piece of land. Here's the bombshell, it is God's land people. And Scripture declares it to be the location to keep our eyes on.

But for now, I'd like to hearken back a few paragraphs to the Word as it is in its' oldest form, the scroll. As I understand, this is still used in synagogue today. It's a scroll not a book. Should this knowledge mean something to us or is it just inconsequential information? Or is it much more important than we are led to believe?

I've never had the privilege of reading one but common sense dictates that reading one would be much different than I am accustomed to. It would be, it seems to me, either a fluid read or completely mind-boggling if you ever lost your place. What it does display is the grueling, methodical accuracy the scribes undertake when copying a scroll. One mistake of one jot or tittle and the whole parchment is torn up and the scribe begins again at GENESIS 1: 1.

There's sacredness to the Hebrew scrolls, with its ancient form of gematria, which surpasses our ability to understand. Being English speaking, English writing man, I can't quite put my finger on it but I sense something's been "lost in translation" or lost in transliteration. Not the Truth of scripture, perhaps just mans' translation of that Truth.

One thing I find highly suspect, which I can't blame on the evolution spouters, is the strange omission of God's name in the biblical version of his Word; which specifies repetitively how we are to both know and call on the name of the LORD< that's a title not a name. Another glaring fact is the name Yahshua changed to Jesus. I can't know whom, why or when this was done; I only know it to cause some to be perplexed! Could this change be of Roman or Greek origin? Could be. I do, however, take solace in the fact that his name was inscribed in Hebrew, Greek and Latin at the stake (JOHN 19: 20), so to me Jesus is a valid rendition f his name.

But the partitioning of scripture, via the press printed bible, is what reveals the encoded possibilities of God's Word so I am left only to conjecture the impact. If not for the chaptering and versing of scripture cross-referencing would be a dubious task at best. The truth is I don't know! I can only go with the information presently available and as that information grows or increases or opens up more, we'll all have to respond accordingly.

As I stated earlier, I believe that, when all is said and done, Holy Scripture could be revealed as visibly hologram-ic in nature, substance and clarity; dealing with every manner of creation; those conceivable, as well as those yet inconceivable by mankind.

What I do know is this; it has in no way been an easy task to determine scriptures' hidden, coded and scientific authorship! God authored the bible through man, just as he makes salvation available through a man, Yahshua Elohim/Lord Jesus! He could have easily, at any moment in times' reality, crushed the head of the serpent, but Satan would forever cry unfair, unfair.

It would have appeared to be a classic case of "big bad" God picking on "little old poor satan", thus grounds for the adversary to cry foul! So Yahweh El Shaddi took on the form of man and defeated satan at the core and the heart of the matter when he defeated him as a man, in the person of the Messiah! YHWH is just and fair and defeated the devil in a just and fair manner and measure! Do you know how? He obeyed Torah. He actually, unlike man, is just and Righteous and Holy.

Friends, it is time for spiritual maturity to replace replacement theology! It is time for spiritual maturity to rise and demolish the mind manipulation of evolutionary trainings! It is high time spiritual maturity replaces psychological justifications! It is time spiritual maturity debunk liberal theology! It is time for our spiritual identities to be realized! It is time we, the true body of Messiah, are in the world but not of the world! And soon it will be time for mans' manipulation of God's creation to come to an abrupt end, for the days of Noah have returned upon us all!

May Father God, Yahweh, Elohim, El Shaddi bless you as you go and bless you as you come according to his good will and pleasure. May we all find forgiveness, I pray, at the foot of the cross-and the precious blood shed there! In Yahshua Jesus name I pray, amen!

For those of you, who truly are believers and not just lip service believers, persevere! For the worldly, proud, doesn't need no God, atheistic evolutionistic manipulation crowd I say this; this book has just affected your future, eternally! How you say?

Accountability by witness! Your excuse of having never heard of the Gospel is gone! You will not have the leisure of claiming, "we didn't know". If, however, this writing in any way can assist you surrendering to Christ, the angels in the heavens, not "space", shout with joy! If you ignore it, your crutch of "excuse" is forever gone! Welcome to the power of the double-edged sword!

For the "lip-service", psychologically tuned, liberal theology believers, scripture does announce your future so I'll not expound. I will say this; you need to open your bible right now and read ISAIAH 29: 13-16 and MATTHEW 15: 8!

And before leaping to conclusions, know this; my heart yearns that all of mankind finds their way home to God! God created the heavens and the earth and all that is in them, which includes you too Mr. evolution, so "Repent for the kingdom of heaven draws near" (MATTHEW 3: 2 and 4: 17). I know, "you can lead a horse to water but you can't make him drink", right? Well, I have a caveat to add; you can absolutely make him thirsty on the way, though! And read what Christ says in MATTHEW 5: 6 and JOHN 4: 14.

Now, continuing with the flavor of this speculation chapter, I'd like to touch on a couple more topics before we bring this speculation chapter to an end. The "rapture", what will that day be like?

This is purely conjectured on my behalf, but I think it's possible that this description of the "dumb masses" being "caught up" moment is different than most people realize. Let's see what scripture declares. First read 1 THESSALONIANS 4: 15-17 for context. Now that we have reference, cross reference that with what Christ states about his return in

MATTHEW 24: 27 and 24: 40-41, and in LUKE 17: 22-37.

I want to pay particular attention to Christ's' description of vultures and carcasses, which will mark this time. It's much different than the church's thinking on this matter. The church teaches a mass, skyward exodus, consisting of both body and spirit being "raptured" away in the twinkling of an eye, which would leave non-believers bewildered with confusion asking, "hey, where'd everybody go".

First) I think the church thinks many more than are actually going will go. Second) I think the church thinks it will include their physical bodies. I think the church discounts scriptures' description of this future event. Third) there has been too much guessing as to when this will occur, so let's deal with them one at a time, ok?

First) scripture declares in MATTHEW 7: 14 "small is the gate and narrow the road that leads to life, and only a few find it". This, alone, should strike fear in the heart of every man, woman and child alive. And it should force us all to drop to our individual knees and beg for mercy, not take it for granted as many in today's society do! Remember the groveling taking place today, as many seek to become American idolized? They grovel at the feet of these judges well enough but what of striving to enter eternal glory?

Listen, get to the gate first or get to the gate last, I do not care, just strive to get to the gate! I have no power or authority to promise entry through the gate, but I know striving to get there is necessary. If you have no care in the matter, why should the one who does possess the power and authority to allow entry, care?

Second) Yahshua says point blank that there will be carcasses and vultures accompanying this "rapture". That tells me that there will be many dead bodies lying right where their spirits left them, not taken up as many like to hope. I don't foresee many people disappearing at one time, I see many people dropping dead at one time for no apparent reason. Talk about bewildered confusion! Remember, the only way for the "perishable" (flesh) to clothe itself with the "imperishable" (spirit)" (1 CORINTHIANS 15: 50-54) is when the physical body is eliminated.

Third) timelines create a paradox! Nobody knows the day or hour, period! Yes we are charged to "keep watch" and read the signs but only the Creator knows when the created has reached its' point of conclusion! Pre-tribulation rapture, post-tribulation rapture, mid-tribulation rapture? Who knows other than the one in control of it all? What I do know, thanks to a growing body of Torah keeping gentiles and a growing body of messianic Jews, is that it is intricately linked with the culmination of the fall feasts of the LORD. A cross-reference of scripture seems to indicate both pre and post tribulation rapture. The truth is we don't know. It is a day and hour that no man knows.

What we can know, however, is if we are living in obedience to his Word, or repentance for our disobedience. In the end that is the one thing we can really know! If you have assurance of salvation yet fail in either obedience or repentance, you may want to re-evaluate your relationship with Christ. Pay very close attention to the words you speak!

The last speculative topic I'd like to discuss is caveman and dinosaurs. When did they walk the earth? Science has its' theories but what does the bible say? We're talking about the stereotypical Neanderthal, grunting, dumb man and massive creatures that never would have fit onto Noah's ark.

Let's begin with the caveman. Could this be descriptive of pre-fallen man, dumb to the things of knowledge? They've been portrayed differently through entertainment and education but is it possible that they were, in fact, wiser than we are? They had Wisdom whereas we have knowledge? Could Adam and Eve have been caveman?

It may seem like a silly idea to float but weren't they, caveman, void of knowledge, language, mathematics and intellect? That's how they are portrayed; "dumb as an ox", grunters, hunter gatherers and void of any knowledge of good and evil, at least until they disobeyed and "Then the eyes of both of them were opened, and they realized they were naked" (GENESIS 3: 7)? Could this describe the "psychological" addition of the "sub con**science**" to the con**science** mind, or vice versa? Before heaping condemnation, remember I'm merely speculating, so let's speculate some more.

This description of eyes opening is also eerily reminiscent of a youth becoming a teen. I state none of what I say in this chapter as fact or interpretation; it is merely speculation-not provable, not un-provable! In truth it's not even worth speculating, I just threw it in there based on the present topics of discussion and the radical change we all under-go when puberty dawns on us all.

What of behemoth with his tail, which sways like a cedar? A cedar is a big tree people. Or what of his bones that are as tubes of bronze? What of leviathan? Read JOB chapters 38-41 in their entirety, as this description fit that of "dinosaurs"!

My fellow man, time is short, speculation is everywhere, truth is "subjective" and secularism is adored. Stop! Take a deep breath; determine your hunger and thirst and proceed with caution into the days ahead as they unfold. Do you hunger and thirst carnal desires or spiritual desires? To satisfy the carnal you have the **world** at your service, albeit temporal. To satisfy the spirit you have the **Word** to offer your service to.

Black and white, there are no "shades of gray" with God. Hot or cold, no luke warm, as REVELATION 3: 15-16 declares "I am about to spit you out of my mouth". Messiah himself spoke these words so we all better take note and respond accordingly. This is to those within the church, those bent toward theologies; as the evolutionists, those bent toward theories, don't give thought to what Christ declares!

I purposely enlisted my own chapter of speculation to make a point; it's pointless to speculate! We either believe or we don't. We obey or we don't. We repent or we don't. Black or white, hot or cold, right or wrong, good or evil, world or Word, God or man! Which side of the fence will you reside on? Neither theories nor theologies are scriptural, whereas obedience and repentance are!

My purpose for this book is not to expound or add to God's Holy, perfect scripture. It's not to proclaim a new way to salvation. It's not to insult, injure, harm or divide the church. It is only an attempt to shake us out of our "comfort zones" of evolutions' methodical intravenous drip. It's to jog us out of our complacent, brick-like, psychological coma, to the realization that we are being manipulated, deceived, lied to and led astray.

Man has elevated himself to the loftiest of heights with his stolen, temporary, creation manipulation knowledge! We stole it from the garden and are as guilty as Adam and Eve were! We've trekked through periods of antediluvian man, the patriarchal period, the given Torah, and the settlement in the Promised Land!

We've had the dispersion of that settlement of people, a recall of the remnant, the Passover fulfilled once and for all, the resurrection of that sacrificial Lamb of God, the times of the gentiles embarked, the sealed scrolls unsealed, the dawn of enlightenment, fulfillment of prophecy and the birth of modern man.

The sealed scroll is almost completely unrolled, time dwindles, knowledge increases, the manipulation of creation continues, mans' heart hardens, good is called evil as evil is called good, God is denied and soon a false messiah will emerge. His spirit and his anti-Christ message already exists cloaked and shrouded in the form of Freud's, Pavlov's and Darwin's theories! Spirit of anti-Christ didn't begin with these three but their teachings have fueled the inferno.

Dark, "**days of Noah**" are upon us as tribulation lies just ahead and the imminent return of him who made us approaches. Nothing we've created will last and nothing of our modern world is worthy of our trust! "Repent for the kingdom of heaven draws near"!!!

Just remember my fellow rock dweller; it is all God's! His creation, by his measurements, his Word that spoke it all into existence and his decision as to what to do with it all! Speech is his, calculation is his, music is his, and electricity (lightning) is his JOB 38: 24, scripture is his, science is his and life and breath is his! What is not his? The theory of evolution! What is not his? Psychology. What is not his? Liberal theology. These all represent the spirit of anti-Christ! These are man's teachings, not God's!

My hope is that each and every one of us wakes up to the spiritual battles that rage within, around and for us all! It is real, it is present and it is really time to fight back against the trainings of the wicked! May YHWH, Father God, Yahweh El Shaddi, through his righteous Holy Spirit guide you in to his absolute Truth!

Through the shed blood of Christ, Messiah Yahshua Lethem, Lord Jesus, amen!

Chapter 10: Hitting Home

Okay, I had to do it; I just felt compelled to add one last chapter before bringing my ranting tirade to a close. If you haven't already closed this book and run in the opposite direction, this might be the point at which you do. This is where the proverbial dog chasing its' tail becomes evident; where the merry-go-round of life gets scrutinized.

This is where the new house, shiny car in the driveway, white picket fence, credit card wielding, two and a half children syndrome, yes the "American dream", is exposed. Like all things and thoughts generated by man this too, started out well enough but look at the fruits they have brought to bear! This whole "dream" has been, and is still being, "bailed out" by the American government. Look at the industries that the "American dream" spawns.

I believe the "American Dream" to be a good dream of freedom, self-accountability and the ability and desire to serve our fellow countrymen while carving out our own niche. What I loathe is the progressive mindset of entitlement and "social justice". I don't believe the government should be the big, expansive dictator it is becoming. I don't believe culture should influence us the way it does. I don't believe our judicial system should be doling out laws from the bench.

I don't believe the American Dream should morph into the one-world touting entity the progressive new-agers so desperately desire. I don't believe the American Dream should become the kum-byah, one world, government is god, hence Nimrod at Babel, nightmare the dream is becoming! I just don't believe the American Dream, at its' core, should seek to rule the world, through the collapse of the American Dream. That evolutionistic thinking only leads to nanny-state governence! How?

What of the housing industry? Bailed out. What of the auto industry? Bailed out. And what of the banking industry that makes mortgaging and financing anything possible? Bailed out, all government entities now. Besides, anything mortgaged or financed is just a means of acquiring now, that which we can't afford now! We have a bucket with a gazillion little holes as our only means of shedding the water (debt) we're taking on, as scruffy chases the tail! Do we really long to be indebted to China, or the world, for that matter?

Is commun-apitalism what our founding fathers envisioned? How about capit-ommunism? Well, on the great chessboard of life that is exactly where we are headed if we're not already there! Can you say mark of the beast microchip, checkmate? It's already done to animals *and* is offered up as a possible child protector too. I guarantee we will hear of this more and more as time passes! How and why? Re-read the four preceding paragraphs again.

I fear it will also be the ultimate fruit and cost of the American dream. I don't want this; do you fellow American? America's Manasseh or Ephraim identity has been chipped away, both collectively and individually, to the point of no return. We are now mere citizens of the UN run, Agenda 21 pushing, "global community". A "one world" happy little family if you will. What do you know? This too is scripture being fulfilled right before our very eyes; you know the eyes God created and man strives to sway us in to believing otherwise!

Did you catch the implication? If you are of European descent or American through

European descent, you are quite probably of the tribe of Manasseh or Ephraim, thus Joseph, thus Jacob, thus Israel! Hear, O house of Israel, Shema Israel! Do you now understand the importance of knowing our true identities? Know who you are as "the days are evil" EPHESIANS (5: 15-17). **NOTE**: even if we aren't blood-linked to the House of Israel we are, still, believing gentiles, thus, grafted in to, therefore, considered part of, the House of Israel.

And what yet lies ahead? A "one world government", a "one world monetary system", a "one world military" and a "one world religious" apparatus are already forthcoming. Honestly, do you think "**social security**" is not the beginnings of **social**ism? And rest assured, there certainly is no **security** in a government run program! And "universal health care" will be the dictated requirement of all. A world implementation on this issue is quickly approaching. And it will permanently change and "fundamentally transform" our country!

This is no conspiracy either! It is the sad reality of the evolution drenched; mentally manipulated, ripe for the taking, thoroughly modernized state we have been manipulated to.

This is, for lack of a better term, the "chickens coming home to roost"! This is the "monster" that we've "created"! This is the epitome of "sowing to the wind and reaping the whirlwind". This is nothing more than "monkey business". This has been their goal all along. The question then becomes; what do we do about it?

1) Scripture foretold this would happen. What, did you forget to believe that part? 2) Scripture foretells that it's actually going to become much worse. What, did you pass it off as merely "figurative"? And 3) scripture declares that no one will be able to buy or sell, save them that bear the mark. In fact, it states that death awaits you if you deny this mark! I can't know for sure what this "mark" will entail, but my gut says it will be some form of implant, so let's break it down.

1) Thirty-years ago, when I was but a young man, none of this seemed possible for generations to come, but that no longer applies. I remember reading the bible, looking around and thinking; "how will persecution be relevant in a free country like America"? It was the 80's; technology was a cassette tape and microwave. A remote control was introduced to "surf"

multiplying channels as cable TV gained steam after the advent of MTV™, HBO™ and the like. Pre-cable news networks, pre-cell phone and pre-internet; you know, the Internet Al Gore theoretically "invented". Oh if I had only known back then what I know now.

But a mere three decades later and look at us. The progressive movement moves along full steam ahead! The sixties hippies are all growed up and in control of our country, with their "higher than mighty" ideology! Their pro abortion, prayer-less education, evolution soaked, same sex marriages, anti-war, "peace"; no ten commandments displayed publicly, activist legislature rules the land! Now, persecution is a certainty! For now it is being held at bay as verbal assaults through the secular thumpers, but just over the horizon, those insults will shift to active injury and death.

2) It will get "**progressive**ly" worse for those who dare to trust God over man. They think, and really believe, we are a bunch of "dumb mass" monkeys; they being the ones in the "know"! Just ask Bill Maher, Janeane Garofalo, Sean Penn, Melanie Griffin or Ted Turner if your measly little, unwanted, scriptural

based opinion matters! If only they possessed one inkling of insight to back up their venom spewing falsehood. But hey, that's what they are good at, calling people names and verbally assaulting anyone not of their mindset! They hate those who "cling to their guns and bibles".

And what do these, and those like-minded, have in common? They're filthy rich, unforgivingly bias, not God fearing, mean, nasty and are given a forum to spew their arrogant venom! For some uncanny reason the media monsters highlight their narcissistic, self indulgent, shallow, petty, vain, irrelevant opinions!

The "beast" feeds on their folly, but again, "The tongue has the power of life and death" (PROVERBS 18: 21). And remember; "For by your words you will be acquitted, and by your words you will be condemned" MATTHEW (12: 37). They're speaking themselves into oblivion so let them!

Scripture declares, "If you are wise, your wisdom will reward you; if you are a mocker, you alone will suffer" (PROVERBS 9: 12). They will suffer for their mocking so remain strong fellow "dumb mass", "idiot" labeled recipient!

And take solace in what scripture states in PSALMS 102: 18-28, regardless of *their* standing and position as satan's mouthpieces! I can't know how much alone time they spend weeping, but their public "gnashing of teeth" is on display for all to see; which is exactly how their needy, boo-boo lip, desperate for attention lives need it to be!

And as our country slides into their notion of a utopia granted global community, they all will progressively rejoice at your injuring, fellow brethren believers! They hate Christ, they hate you, they hate me, and they hate anyone not of their thinking! Do you want to know how to make their blood boil? It's a dirty little word, are you ready? Prayer! That's right, pray for them! And heap the burning coals back onto their heads! Ok, that's about enough ink wasted on them so let's move on to "the mark", a deed they will be more than willing to both endorse and partake of.

3) This is the height of "days of Noah" Nimrod brutish man control and it could very well cost you your life. Scripture states that if you receive this mark you're done, check mate, finito, over, no mas, done! If you or I, in any

way, cherish or treasure our earthly possessions, hence the materialistic playground the American dream affords, doom beckons! This is not I speaking; it's God's Word that says such things. And as I've alluded to frequently, we had better take scripture literally, accept for when it specifies otherwise!

For clarification, let's take a look in the book of REVELATION. Specifically, REVELATION 13: 1-18 states what lies ahead. And REVELATION 14: 11-12 reiterates the call to reject this forced mark. It should be quite evident that this is a real scenario described in God's Holy Word, so resolve within your being to prepare for the arrival of and rejection of this, I believe, computer microchip implantation!

The harvest is ripe and the axe is at the root of the tree. The stage is set for the man of perdition to arrive. I don't know whom or when, I only know scripture declares it, so it will be! Evolve that thought away Mr. atheist evolutionist! The horn is being raised, the mighty breath being inhaled and the trumpets ready to be sounded, people, so "get your house in order", "get your priorities straightened out" and humble yourselves!

God's Word is quite clear on this issue; you humble you or prepare to be humbled, by the very hand of God. Clear away the smoke and mirrors, drop the scales, lose the blinders, demystify your thinking of the glitz and glam culture which currently hypnotizes us; seek God and start living as though the kingdom of God were already set up here on earth! Strive to live the kingdom way of life as to separate from the "worlds" way of living; and when the man of perdition arrives you won't fall for his lies.

Do we think we can live as the world does then, presto, be handed the keys to the kingdom when the transformation occurs? No, the invite is to those sanctified, set apart; who know, and prove it with speech and lifestyle, a different standard of living! Trust this; evolution trainers and their astute valedictorian students aren't invited unless repentance super-cedes their notions of our Creator! And if you didn't catch what I just said, yes, we will have to prove ourselves worthy to enter the kingdom's gates. Not with works, by faith.

And if you foolishly say in your heart: "I don't want a God that I have to prove myself to", good! That is exactly what you will get, no God!

Forever! Be that as it may, if you do cling to the very Word of God, a quality held with complete disdain by our current president, good, I look forward to meeting you in eternity!

Sort of in that vein, I'd like to ask a question that has perplexed me for a while now. It's about those who seem to think that "aliens" had something to do with our being created; the "chariots of the gods", crowd. They look anxiously forward to this being revealed, why? They harbor bitterness toward The Living God and ask things like "how could a loving God leave us here to destroy ourselves"? "Why is there murder, kidnapping, child molestation", "why are there disasters"? Yada, yada, blah, blah! "Why"? The answer lies in a little word that they've been trying, for decades, to omit from our vocabulary; SIN! "Why do bad things happen to good people"? Sin!!

Well, their questioning presupposes two erroneous thoughts; first, that people are basically good, contrary to what scripture declares. And secondly it presupposes that what we determine as bad is, in fact, bad! Let me give an example: your child is playing catch and the ball heads into the street. Your child darts straight into the street, not knowing that a car is quickly approaching! You yell, while sprinting toward your child, no! Stop right there!

This child, because they neither know as much as you, their parent, nor do they perceive the approaching vehicle, "feels" angst that you prevented them from getting their ball, as the oncoming car punts it down the street. They perceive it as "bad", whereas you just didn't want to witness your beloved child being obliterated by a car. So what we, children of God, determine as bad, might not necessarily be an accurate description of the situation!

Now, concerning the "aliens placed us here" and "altered our DNA" crowd, why the double standard? Seriously, why the double standard? Why no bitterness pointed toward the "aliens" equal to the squalor pointed toward my God? If you charge God as guilty, and you know you do with your "why does God let bad things happen to us", then why no guilt pointed toward your precious, imagined aliens of origin? Seems to me, dare I say it, hypocritical!

I surmise this conjecture by one Erich Von Daniken as being conceptually right yet backwards. He hypothesizes the "gods" of the ancients were actually "aliens", whereas I surmise that the "ancient aliens" were in fact fallen angels and demons, posing as "gods". Either way, why the anger at God for our current state, yet no angst toward their "aliens" for our current state? I'll not tell this man or any of his followers what to believe, but I will point out the hypocrisy of giving one, "aliens" a free pass while the other, Creator God, is mercilessly scrutinized beyond measure. Moving on,

I feel compelled to share something with you that I had no intentions of revealing. What's that? Well, you are reading a "dead" man's writing right now, at least according to those in the "know"! Come with me on a journey into the realm of spiritual reality and pay close attention to what I'm about to share with you. This is, "figuratively", my ball shooting across the street metaphor.

It was a beautiful, sunny morning on May 23rd of 2005 when the stark reality of spirit meets flesh occurred to and for me! I'm a carpenter by trade and that Monday seemed as any other workday, as we were preparing to shingle the roof we had just refurbished the previous Friday. We were rolling out the felt paper, right over the three skylight openings spaced equally across the roof.

In what seemed like a dance with weightlessness, I stepped right on, hence through, one of the tarpaper covered skylight openings. I knew better and regardless, still *fell* victim to my own confidence. I will never forget what followed, and I'm sure those whom it had an immediate effect on, will never forget the day of "thud", as my dad tagged it, either.

I admit it; I was a knuckle headed bonehead! I fell twenty-three feet in to an empty, indoor, in-ground, concrete pool. Snap an ankle or two? Nope! Shred a knee or two? Nope! Break a leg or my neck or my back? Nope! I shattered my forehead and left wrist, whip-lashed my back and neck and had blood on my spine. I recall telling the boss to just put some duct tape on my wrist so I could get back to work, not knowing the damage done to my skull.

I didn't crack or fracture my skull; I shattered it completely! And I did it to my frontal lobe. Go ask any doctor you can find how vital the frontal cortex is to basic human function. The nurses in the ER had me dead by nightfall; the emergency room staff had me dead by nightfall. The doctors and surgeons had me dead, paralyzed, comatose or in a permanent vegetable state. My prognosis, after these four possible outcomes were miraculously eliminated, was to suffer seizures and suffer irreparable damage to both my short term and long term memory faculties.

The surgeon, who pieced my fragmented head back together, after having to expedite the surgery due to a rapidly growing blood-clot on my brain, told me that he has personally witnessed three people drop dead, on the spot, with lesser injuries to the same exact area of their brain! My wife told me of a recent story on the news where a man fell twenty feet through a skylight and broke one arm and one leg……… Then died!

But there's more! I never even lost consciousness; not once did I black out! I would like you to picture a hardboiled egg falling off a standard height kitchen countertop, smashing to the concrete floor below, shattering the outer shell and yet the membrane which surrounds the hard boiled part of the egg remaining completely in-tact. Doesn't seem possible, does it?

Well that is precisely what happened to me that day! A mere pinhole sized tear in the membrane that surrounds our brains would have rendered me incapacitated completely, with one of the four outcomes aforementioned as a done deal! On top of that was my shattered left wrist, which needed to be surgically repaired with pins. Until, of course, the next morning when the surgeon informed me that my wrist had some how went back together on its' own, praise YHWH, The Living God! My wrist was still broken, just not to the point of needing pins anymore.

I do not know if it was Christ himself or one of his Holy angels with me that day of "thud", but I know something miraculous and scientifically unexplainable had happened to me. How many seizures or blackouts to date? Notta! Amount of memory loss suffered to date? I can't remember; just kidding, the answer is miraculously zilch! Amount of days spent in the hospital? A mere five! How many observable residual effects to date? Oddly enough it's the, at the time unimportant, effects of back and whip-lashed neck pain that remains. I presently possess four bulging discs and five arthritic vertebrae in my lower spine. So when I say trust God over man, I whole-heartedly mean it! And with good reason!

I do feel as though I need to clear something about that incident up too. There were many who prayed fervently for me following the prognosis doled out from the surgeons, doctors and nurses on staff that day. I am forever grateful for their prayers, yet I take offense to those who attribute my life being preserved to *their* ability to pray, a notion that confronted me upon my return to work seven months later. The prayers didn't begin until after I had reached the ER, not before my head slammed into the pavement, which is the moment I was miraculously spared!

Again, I am forever grateful for the outpouring of prayer and support, I just don't like when people take credit for a miracle by the hand of God! My life was not only spared but also completely preserved the moment my skull made introduction with the concrete, way before the first prayer was even uttered. I give God all the credit for me still being here, not man! If this seems insensitive, hearken back to the hard-boiled egg scenario.

I already told you of the schwannoma tumor I recently had removed. This too, is a rarity amongst my human counterparts, with only one in millions acquiring such a fate. One elderly doctor told me, pre-op, that he hadn't heard of such a thing in thirty some odd years. Actually, I've heard more than one similar comment from more than one doctor.

And as I scan the memory banks, the ones God fully preserved, there are a couple of other scenarios, which should have left me dead and buried a long time ago, but they didn't! One was a fall from a second floor top-plate, through 2 stairwell openings and to the basements' concrete floor. Another was a car accident in which, as a passenger, a guardrail plunged right through the passenger side door and crumpled the seat I was occupying. I'll spare the details but know this; when I speak of God, his Word and the truths of life, I choose my words carefully, speak them purposely and mean them wholeheartedly!

I say what I mean and I mean what I say! God is real, fellow man! God's Word is real, fellow man! Evolution is a lie, fellow man! Psychology is misleading, fellow man! Liberal theology is deception, fellow man; and this will all be sorted out upon Christ's return, fellow man! Come May of 2005, I had set *my* goals when God showed me *his* will, a painful yet necessary lesson! Why? I wanted to shift from employee to small business owner.

Eight days prior to this day of "thud" experience, my wife and I had put our house on the market and had plans of relocation. If you notice the year, it was smack dab in the middle of the housing boom; we would have walked away with a pretty substantial profit. Against my instincts and based solely on physical desire and my red blooded "American dreams" of grandeur, I had set a goal when God showed me his will.

Once fully recovered, I peered back and discovered God actually protecting my family and me. You see, I had already begun the ball rolling on a no doc.-arm mortgage, which would have, unknown to us at the time, been financial suicide. We would not have made it through that kind of financial dealing, as it's not scriptural! So the day of thud was, from a spiritual view: a special, multi-pronged miracle given to me by our maker. It was a lesson I had to learn. Why?

Well, I had long suspected the spiritual realm interacted with and influenced our lives more than we're aware of, never really knowing the why or how such things occurred. After "thud" day, and not knowing why I was spared that day- a question yet to be answered, an insatiable appetite for God and the spiritual things of God developed in me. I began asking different questions, I began a long forgotten quest for truth!

But the Truth I seek is not of this world! I seek God's Truth, not the speculative, "subjective" truth man offers. That kind of thinking just won't cut it anymore! It is only recently; as I hearken myself back three decades that I realize the effectiveness and success of evolutions' trainings, and the necessity to finally take a stand against such ill flavored ilk!

This phase of my journey kicked off in 2005, yet has sped up exponentially since the tumor reared its' life altering head! It wasn't until this tumor knocked me off my feet that my eyes were opened to the truth that science is, in fact God's. It is of his wonderful, all surpassing, glorious Wisdom; praise Yahweh! And what I have learned through these two life-altering experiences is what I have attempted to share with you via this writing you currently possess.

To put it bluntly, I stopped chasing my tail; where "thud" got my attention, schwannoma produced commitment! Like I said: humble yourself or you will be humbled! I spent the summer of 05 bonding with my three oldest children; my daughter Paige and my two oldest sons Zechariah and Michael, post thud. I have had the blessing of the birth of my two youngest sons Jeremiah and Isaiah, post thud. I've learned of the creative powers of the Creator in a different light, post thud. I have experienced the hand of God upon my life on different levels, post thud. I know it was always there, I just hadn't been aware of it to the extent I am now; too busy chasing the "American dream"!

So, my fellow 3rd rock dwelling man, forgive me if I seem a tad overbearing in my approach. I have just reached a point in my life where I will not sit quietly by and allow my own kids to be pummeled by the filth doled out to our children through the stealth like subtleness of psychologically endorsed evolutionistic trainings; nor will I ever willingly attend a church where liberal theology abounds! I will tell them, my children, the Truth of scripture, for there is too much at stake!

Remember, I could have very easily tagged these experiences of mine as "bad" experiences, much like the child chasing the ball into the street; but as I said, that would not have been an accurate assessment of the situation. No it didn't "feel" good at the time, and its' lingering effects are still prevalent stress inducers, yet it sure has bore some fruit that I may have never become aware of were it not for these seemingly "bad" things happening. Personally, I believe Elohim is, and has been for a good while now, preparing me for the days to come.

I must stipulate that it is with a broken heart that I proceed. The reason? Two weeks ago I buried my Dad, my hero; I watched him breathe his last. I have never felt such a deep pain, nor experienced such a deep sorrow. His passing is having a deeply profound effect upon my wife, my children and me, as the loss is heavy. So as we grieve, I state emphatically, wake up fellow man. Wake up and turn to God before your God-given, God-ordained last breath is breathed. Just so you know, I did check his hands and as I suspected all along, the man did, in fact, have precision/power thumbs. Do you know why? He wasn't a "dumb mass" monkey. He is my Dad! Got it Mr. Evolution?

Okay, let's quickly clear an issue up before concluding; dividing the Word correctly is the opposite of separating the Word. As I stated earlier the Old and New Testaments, or Covenants, are single cohesive bodies. Not to be separated, not open for interpretation. Dividing the Word correctly means harmonizing Old Testament prophesies with New Testament fulfillment and harmonizing New Testament prophecy with what is being fulfilled in our "age"; by observing Old Testament times and seasons.

Dividing correctly is harmonizing scripture as it is. Knowing when Christ spoke as God and when Christ spoke as man, knowing the difference between the spirit and the physical. Dividing correctly is to "Give to Caesar what is Caesar's, and to God what is God's" (MATTHEW 22: 21). Interpretation is Holy Spirits' duty, dividing correctly is mans' duty and separating the two Covenants is nobody's duty.

Well, here we have arrived at the **mock** through **epalt** leading to idolatry. The **M**anipulation **O**f **C**reation **K**nowledge through: **E**volution, **P**sychology **A**nd **L**iberal **T**heology leads to gross idolatry, so be informed fellow man! Just avoid theories and theologies altogether and cling to and strive to obey scripture! We will fail at perfection until Christ fulfills it in us-which we can't control! But we are commanded to strive toward this end regardless, that we can control!

So, as the "worldly", "global community" internationalizes us unto one big happy global family, I will respectfully decline that invite for I seek a better promise, the Eternal one! Praise God, praise his Holy and Righteous name-YHWH and praise be to his Holy, timeless, always right and just, Wisdom filled **Word**! Thank you God, for your Word to us through the bible and your Holy, Righteous Son Yahshua Jesus, amen!

CHAPTER 101/2: MY FINAL PLEA

Well my friends, this is my final appeal to your senses. Do you speak ooh ooh ooh, or do you speak language? Do you recognize aah aah aah, or are such notions insulting to you? We have a choice to make. We have a decision to formulate. We have a conclusion to reach regarding this issue.

There is no in-between; there is no gray shade here. You either buy into the lie of evolution or you don't. You either believe in slime metamorphosis or you don't. You either came to be by transmutation or you were created. You are either of the world or of the Word. "Do not be misled: 'Bad company corrupts good character'" (1CORINTHIANS 15: 33)

If you are able to decipher oooh oooh oooh – aaah aaah aaah, then maybe you are the missing link, I jest. But I think otherwise. I think something deep down inside of you knows, just knows, that something is off kilter about this evolution gobbledygook.

I surmise, with all the hyperbole I can muster, that we all know better! I contend that we all know in our gut that Darwin's death-bed confession rings truer than his theory of evolution ever rang! I believe we all recognize our deplorable guilt of eating this rank fruit. If not; we should.

But here is where I do an about-face and declare that there is hope. Granted, it's a fleeting, slipping through our fingers hope, but I'll take it. I do believe that as long as that big gold disc rises every morning, there is still time; there is still hope!

We can do this, man! We can smelt until the dross separates. We can rid ourselves of false doctrine and ready ourselves for the days which lie ahead. In fact, we must. Time is much shorter now then when you began this read.

Know that for those who both teach and believe evolution to be truth, they will one day be confronted by the maker they so stringently deny. But for those who are "Christians", "the church", the "body of messiah"or"believers", I offer you scriptures view on the matter.

'For the message of the cross is foolishness to those who are perishing, but to us who are being saved it is the power of God. For it is written:

"I will destroy the wisdom of the wise; the intelligence of the intelligent I will frustrate."

Where is the wise man? Where is the scholar? Where is the philosopher of this age? Has not God made foolish the wisdom of the world? For since in the wisdom of God the world through its wisdom did not know him, God was pleased through the foolishness of what was preached to save those who believe (1 CORINTHIANS 1: 18-21).'

"The man without the Spirit does not accept the things that come from the Spirit of God, for they are foolishness to him, and he cannot understand them, because they are spiritually discerned (1 CORINTHIANS 2: 14).

I find myself a bit perplexed that this topic of evolution is a topic which even needs addressing; for the world and, certainly, for those in the church. I'm befuddled to think that "teachers" could teach such a thing; but I can't fathom that Christ's bride would even consider believing such a teaching.

What follows is my final attempt to jog your reason. My hope is that any and every set of eyes reading this right now just soak it in. In the end it is better that we all put away our opinions, restrain our desire to speak and simply listen; and hear what is being laid out before us. Pay very close attention.

This, my brethren, is a wise mans' admonition:

"Dear friends, this is now my second letter to you. I have written both of them as reminders to stimulate you to wholesome thinking. 2) I want you to recall the words spoken in the past by the holy prophets and the command given by our Lord and Savior through your apostles.

3) First of all, you must understand that in the last days scoffers will come, scoffing and following their own evil desires.

4) They will say, 'where is this "coming" he promised? Ever since our fathers died, everything goes on as it has since the beginning of creation.' 5) But they deliberately forget that long ago by God's word the heavens existed and

the earth was formed out of water and by water. 6) By these waters also the world of that time was deluged and destroyed. 7) By the same word the present heavens and earth are reserved for fire, being kept for the Day of Judgment and destruction of ungodly men.

8) But do not forget this one thing, dear friends: With the Lord a day is like a thousand years, and a thousand years are like a day. 9) The Lord is not slow in keeping his promise, as some understand slowness. He is patient with you, not wanting any one to perish, but everyone to come to repentance.

10) But the day of the Lord will come like a thief. The heavens will disappear with a roar; the elements will be destroyed by fire, and the earth and everything in it will be laid bare.

11) Since everything will be destroyed in this way, what kind of people ought you to be? You ought to live holy and godly lives 12) as you look forward to the day of God and speed its coming. That day will bring about the destruction of the heavens by fire, and the elements will melt in the heat. 13) But in keeping with his promise we are looking forward

to a new heaven and a new earth, the home of righteousness.

14) So then, dear friends, since you are looking forward to this, make every effort to be found spotless, blameless and at peace with him.
15) Bear in mind that our Lord's patience means salvation, just as our dear brother Paul also wrote you with the wisdom that God gave him.
16) He writes the same way in all his letters, speaking in them of these matters. His letters contain some things that are hard to understand, which ignorant and unstable people distort, as they do the other Scriptures, to their own destruction.

17) Therefore, dear friends, since you already know this, be on your guard so that you may not be carried away by the error of lawless men and fall from your secure position. 18) But grow in the grace and knowledge of our Lord and Savior Jesus Christ. To him be glory both now and forever! Amen (2 PETER chapter 3).

Is that not absolutely pertinent to the discussion we've been having? Is that not absolutely relevant to today? The evolution pushing, nay-say crowd claim the bible to be out dated in today's day and age; but I think not! And this is the last little tidbit I'll offer.

First I would like to quote this scripture: "An oracle concerning Damascus: 'See, Damascus will no longer be a city but will become a heap of ruins" (ISAIAH 17: 1). This is time sensitive and a very real possibility today, 8-28-2013.

I realize that I just dated this writ, but we are on the verge of attacking Syria for use of chemical weapons. I will forgo all the political jargon and get right to the point. It is quite possible that by the time you are reading this, this prophecy has been or is in the process of being fulfilled. Out dated; huh?

If you missed what I just said I'll translate it for you; oooh oooh oooh aaah aaah aaah!

Points to Ponder

"Our Father in heaven, hollowed be thy name", "hollowed be your name"? Point To Ponder (PTP); what is God's name? Many times over scripture teaches us to reverence, respect, and be fearful of and to call upon the name of the Lord, yet doesn't reveal his name. Am I the only one who finds this odd?

Pick any modern, mainstream translation of the Word, and you'll find his name omitted, and substituted with, LORD, Lord or God, very peculiar. God is not his name; it's only one of his many titles. Seriously, scripture is just saturated with the importance of his Holy name, yet silent in revealing that name. I possess a 'TODAY'S PARALLEL BIBLE' with the KJV, NIV, NASB and NLT translations, none of which offer any insight to this question. Our Heavenly Father's name is omitted. Again, why? Adoni, Jehova, Yahweh, Yahveh, I AM, EhYeh-Asher-EhYeh, which is it?

This may surprise you, but none may be the answer; this may surprise you as well, all may be the answer. His name is **הוהי- YHWH**. As far as is detectable, his name consists of four letters: Four ancient, Hebrew letters. And if you think this is a small matter, you are wrong, possibly dead wrong! "

Did you know that one of his commands is to know his name? Did you know that another of his commands is to call upon his name? Scripture declares it, so you better seek to know it! It should go without saying, but I am speaking to those in the church here. Do you know the duty of a disciple? It means to learn, to be dedicated to the discipline of learning. Now, if you claim to be a believer you should want to learn these bits of information, especially since scripture teaches them. As a believer you bear the responsibility of knowing whom it is that you believe in. How can you call upon the name of the Lord if you don't know his name?

It is imperative that we each understand that the ancient scribes of the ancient Hebrew text attempted to hide the sacred name of our maker. In doing so, they substituted other names, even hybrids, in its' place. But our Creator never instructed against using his name, he instructed against the mis-use and abuse, of his name in vain, and he forbade the uttering of his name via divination, sorcery, magic and witchcraft.

He never instructed us to not utter his name, he commanded us against any empty, violating or common use of it! His Word actually states we must call upon his name, the name Moses was instructed to tell the Hebrew people upon descending Mt. Sinai. His name is **הוהי**(read < right to left). His name is **היהא-רשא-היהא** (read < right to left), and that's as far as I am willing to go. If you seek to know more, "seek ye first the kingdom of heaven".

Here's another Point to Ponder: much to science's chagrin, scripture flat out tells us that the world (earth) is spherical, so why did man think, up until a few centuries ago, that the world was flat? Did they know otherwise, yet hide it, so as to fabricate a "new" scientific discovery? Here's what the Word teaches, "He sits enthroned above the **circle** of the earth" (Isaiah 40: 22).

PTP: The dynamic number 7; 7 days in a week, 7 feasts of YHWH, 7 planets originally, 7 notes in music, 7 continents, 7 seas, 7 variances of the electro-magnetic light spectrum, 7 colors of the rainbow, 7 parts of the atom, 7 days of creation, 7 years of feast and 7 years of famine in Egypt, 7 revolutions around Jericho, 7 of every kind of clean animal entered the ark and 7 days later the deluge began, 7 unclean spirits that return upon the self-reformed, 7 things detestable to the Lord, 7 falls for a righteous man, 7 women seeking a sir-name, 7 seven's and 62 seven's of Daniel.

7 churches, 7 golden lamp stands, 7 stars or angels of the church, 7 seals, 7 trumpets, 7 thunders speaking, 7 'years' of tribulation, 7 angels with 7 bowls of wrath, 7 last plagues, 7 gates to the soul (2 eyes, 2 ears, 2 nostrils and 1 mouth), 7 scars of Messiah (2 punctured hands, 2 punctured feet, 1 crowning head wound, 1pulp-like back wound and 1 piercing side wound). Remember, no coincidences! This short list doesn't even factor in all of the variances of the number seven sprinkled throughout the written text, i.e. 70- 77 and 7,000.

Wake up man! Open your eyes and arise from your slumber!

- As pertaining to the number **93** and variations of it, found on pages 25 and 26; Jerusalem is declared in scripture to be the inheritance of Israel some **93**0 times. Not 929, not 931, but **93**0: "there is nothing new under the sun", especially co-incidences!

- For the manipu-list, every ailment we suffer now has a medical term attached to it that we, the "dumb masses", are not privy to.
- As does the ingredient label on each consumable product contain its' biochemical name rather than its' common equivalent (plain English).

This one scares me a little, but here goes something; if President Barak Obama is, indeed, the one who forces the mark, a notion his progressively socialistic (communism light) agenda suggests; and America is, in fact, the lost House Of Israel, then we will have, once again, the Egyptians (northern tip of Africa – Obama's motherland) enslaving the tribe of Jacob. Think about it. He's progressively, pro one world everything! I'm not claiming he is anything other than a man, I'm just saying wake up to the possibility. If he's not the one, he's only a short one or two steps removed not because of whom he is personally but because of the position he holds when coupled with his one world ideology, so wake up! I mean, come on, I've heard him touted as the messiah, the savior of the world. I'm looking for ha Mashiach, the Word.

Fellow man, ascending the hurdle of doubting God while believing satan, through man and his false doctrines, is of utmost importance; because what lies beyond this information harnesses dire consequences for the "ignorance is bliss" – oooh, oooh, aaah, aaah monkey mentality.

You must possess the faith to believe God at his Word, with no mingling of seed that evolution requires. In the end it's all about the seed anyway. Physically, spiritually and doctrinally so wake the heck up because you're not a "dumb mass" monkey! Once the implication of being of the House of Jacob sinks in, scripture takes on a completely different dimension. All the apparent bantering back in forth from the "House of Judah" to the "House of Israel", as found in the books of the prophets (ISAIAH-JEREMIAH-ZECHARIAH etc.) is distinguished, you will never see scripture the same. And you will hopefully, once and for all, shed agreement with evolutions lure.

Here is another evidence of the quirky little realities of the evolution herd; I remember, back in the nineties when it was introduced to the world that life was discovered on Mars; remember?

There was a big uproar at the time, and even a public announcement by, then President, Bill Clinton. All of the hoopla was over a fossilized microbe. But here's the blatant inconsistency:

Microscopic remains of some squiggly thing a planet away is deemed life; yet a fetus in a mother's womb here on earth is up for debate.

It is, precisely, that form of inconsistent thinking that lends credence to, and validates my notions of, the monkey speak crowd.

* So as to avoid being accused of hurling unfounded slander I feel it is necessary for me to comment on one Nikola Tesla and one L. Ron Hubbard. I never met either one of these men. My conjecture is based solely in the premise of: "if it looks like a duck and it quacks like a duck".

I neither condemn nor acquit either of these two; I merely am not ignorant of the materials available which shed some light on the finger prints of these men's contributions to our modern world and mind set. I know Tesla denied occult involvement and claimed clairvoyance. I know he supposedly attributed nothing to the spiritual as he deemed everything was mechanical in nature; but flashes of insight, sudden visions, psychic abilities and messages from Mars? The Tesla coil, free energy, radar, robotics, guided missiles, "death rays", weather manipulation, electron beams and knowledge of earth's resonance all point to a man endowed with, dare I say it, celestial knowledge.

Come to your own conclusions, but not until after you have researched it for yourself. As for me; the dots are rather simple to connect.

Concerning Mr. Hubbard (Dianetics) I'll state only this; he was in league with one Mr. "Jack" Parsons (a man steeped in occultism), of JPL – NASA fame; who was in turn in league with one Alister Crowley.

Mr. Crowley was **The** occultist, with ties to British intelligence. You do the math on this one. As for me, I'll simply just, again, connect the dots.

So as we conclude this overwhelming journey allow us to connect one last array of dots; as it concerns every one of us and is laid out in scripture.

"11 For this reason God sends them a powerful delusion so that they will believe the lie 12 and so that all will be condemned who have not believed the Truth but have delighted in wickedness" (2 THESSALONIANS 2: 11-12).

Dot connection 101; what is this strong delusion? I contend it is this Godless modern world forged by the hands of modern man, and all possible science fiction mysticism this modern mindset affords- oooh oooh oooh.

Dot connection 101; what is the lie? The lie is the Godless modern teaching by the mouths of modern man that there is no God- aah aah aah!

H **ELLO TO YOU MY FELLOW MAN**
E **VERYBODY CLIMB ABOARD**

I **NSTANTLY GRATIFIED TO DISTANTLYSATISFIED**
S **PECULATIVE THEORIES ARE ADORED**

R **EMEMBER FOREVER BEFORE IT'S TOO LATE**
I **S A CRY TO MANKIND TO CLEAN THE SLATE**
S **OAKED WITH THE NOTION OF CREATING OUR FATE**
E **VEN THE HEATHEN SHOULD SEEK HEAVEN'S GATE**
N **EVER FORGET WHAT CHRIST'S SHED BLOOD HAS RESTORED**